COLONIAL FORTS
OF THE
Champlain and *Hudson Valleys*

COLONIAL FORTS
—— OF THE ——
Champlain and Hudson Valleys

Sentinels of Wood & Stone

MICHAEL G. LARAMIE

Published by The History Press
Charleston, SC
www.historypress.com

Copyright © 2020 by Michael G. Laramie
All rights reserved

Front cover: Plan of the Fort and Fortress at Crown Point, Anon. *Library of Congress, Geography and Map Division.*

First published 2020

Manufactured in the United States

ISBN 9781467144865

Library of Congress Control Number: 2020932005

Notice: The information in this book is true and complete to the best of our knowledge. It is offered without guarantee on the part of the author or The History Press. The author and The History Press disclaim all liability in connection with the use of this book.

All rights reserved. No part of this book may be reproduced or transmitted in any form whatsoever without prior written permission from the publisher except in the case of brief quotations embodied in critical articles and reviews.

For Nathanael

Contents

Introduction: The Old Invasion Route, 1643–1760 — 9

1. The Richelieu Valley — 21
2. Lake Champlain — 42
3. Lake George and the Upper Hudson Valley — 73
4. The Lower Hudson Valley — 97

Appendix A. Glossary — 109
Appendix B. Documents — 113
   Report on the Frontier Defenses of New York
        by Colonel Wolfgang Romer — 113
   Report on the Harbor of New York
        by Colonel Wolfgang Romer — 115
   Directions to the Commandant at Fort William Henry
        by Captain William Eyre — 117
   Directions to the Commandant at Fort Edward
        by Captain William Eyre — 120
   Remarks on Forts William Henry and Edward
        by Captain Harry Gordon — 122
   State of the Works at Fort Edward
        by Colonel James Montresor — 125
   Memoir on Fort Carillon
        by M. de Pont le Roy, Engineer in Chief — 126
   Remarks on the Situation of Fort Carillon and Its Approaches
        by Captain D'Hughs — 127

Contents

Appendix C. The Legend of Duncan Campbell	133
Appendix D. The Forts Today	135
Notes	137
Bibliography	147
Index	155
About the Author	159

Introduction
THE OLD INVASION ROUTE, 1643-1760

F ew bodies of water in North America have had such a profound influence over the destiny of the continent as the Hudson River, Lake Champlain, Lake George and the Richelieu River. Together, this geographical feature, which also includes a number of larger and smaller tributaries, cuts through the primeval forests of the Northeast and stretches from the St. Lawrence River in the north to New York City in the south. By itself, this would be nothing more than a curiosity, but when one considers that small boats could navigate this route for all but a few dozen miles, it becomes far more important than an interesting product of the last ice age. In fact, even for heavier vessels much of the waterway is navigable. Large segments of this route—from New York City to Albany, from South Bay to St. Jean and from Chambly to the St. Lawrence—are all capable of taking larger vessels. As a case in point, the HMS *Royal George*, a 383-ton, twenty-six-gun frigate, successfully plied the waters between South Bay and Fort St. Jean on the Richelieu River. From Chambly to the St. Lawrence and from New York City to Albany, the deeper waters provided access to even larger vessels.[1]

In an area void of roads and in many cases even simple trails, this natural invasion route had been used for centuries by regional tribes to launch raids against their enemies. At the time of the first French and Dutch colonists, the route was being used by the Iroquois Confederacy in upstate New York to launch raids against their enemies in Canada, and of course, it carried the response south from the Ottawa, Algonquin, Abenaki

Introduction

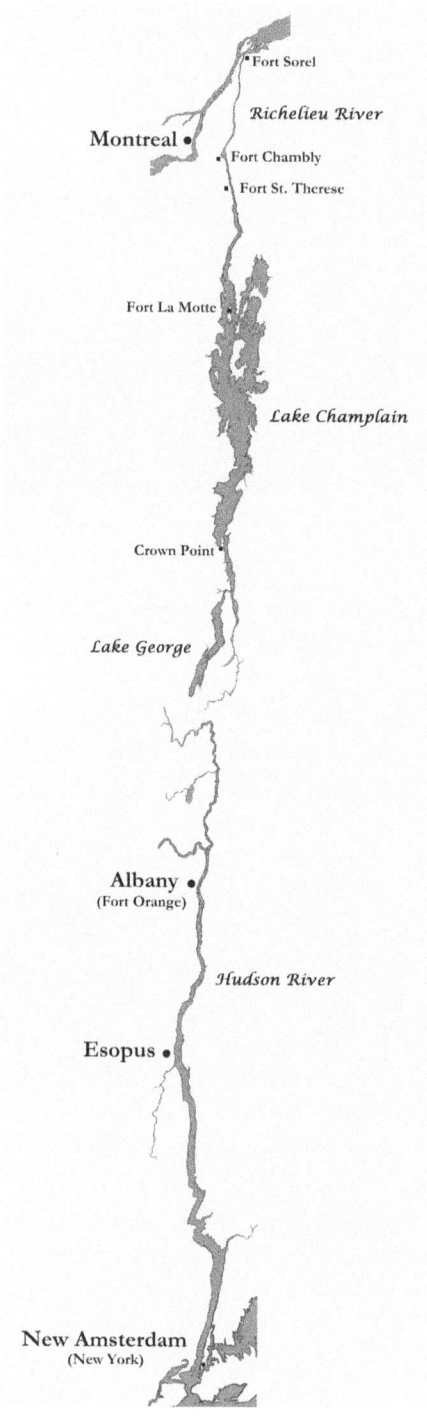

and a host of other northern tribes. It was not until the French established a colony on the north end of this corridor, however, and the English supplanted the short-lived Dutch colony at the southern end that the waterways would take on major military significance. Both sides soon recognized the threat and the opportunities presented by the route and worked to secure their respective ends. The French made the first move to seize a blocking position on Lake Champlain with the establishment of Fort St. Frederic at Crown Point in the mid-1730s. Faced with stone, garrisoned with several dozen troops and armed with a score of cannons mounted in a tower that overlooked a three-hundred-yard-wide portion of Lake Champlain, the fortress effectively barred any army attempting to move down the lake in small boats.

The establishment of Fort St. Frederic gave a decided advantage to the outnumbered French and their native allies. From this stronghold, war parties struck targets as far east as the outskirts of Boston and as far south as New Jersey during King George's War (1745–48).

The Hudson and Champlain Valleys, circa 1670. *Author's collection.*

Introduction

The response from New York and New England was repeated calls for the destruction of the "Tower of Babel," a direct reference to the fort's defining feature. A pair of expeditions were organized in 1746 and again in 1747, but in both cases, an attack on the fortification failed to materialize.

While Fort St. Frederic maintained an undeserved aura of strength throughout much of its history, no fort in North America was designed to be impregnable. This was especially true of the frontier forts. Captain Pierre Pouchot, a French officer serving in New France, perhaps stated this point best, writing in his journal, "One cannot say that the type of fort constructed in those regions is impregnable, given that there is no hope of rapid relief."[2]

In fact, the primary purpose of the fort was to delay the enemy until an expedition could be organized to march to the garrison's aid. While a wooden palisade armed with a few cannons might force the enemy to halt, it would be no match against small cannons that would quickly batter down its walls. Thus, to be effective, the frontier fortresses along the Hudson-Champlain corridor had to be constructed to withstand cannon fire, at least for a short period of time while the alarm was being sounded and a relief force organized.

This requirement was addressed by earthen forts. To construct this type of fort, a giant wooden box, often dozens of feet thick, was made in the form of the structure's outline. This was then filled in with earth, usually from the ditch that was dug at the base of the structure's walls. The front portion of the walls, known as the parapet, was built up several feet in height to provide cover, while the flat rear portion, known as the rampart, provided the garrison and its guns with a firing platform at the top of the wall.

A number of additional elements would soon follow. The bastions, which were essentially blockhouses erected at the junctions of the walls, would be built out to handle cannon and usually served as bomb-proof shelters for the garrison and its supplies during a siege. One of these bastions would also typically serve as the fort's powder magazine. Barracks were then constructed, a well dug, guard houses and observation posts built and a host of other features added, including a network of smaller exterior fortifications designed to slow an attack and provide advanced warning to the defenders.

While the strength of the fortifications was paramount, so was the size of the fort's garrison and their diligence. There were only four ways to take a fort. First, it could be taken by surprise. If the garrison was too small or failed to be attentive, the attackers could be over the walls or through the gates before a defense could be organized. While possible, there were only a few attempts to seize a fort by surprise along the Champlain-Hudson waterways, and far fewer successes. The second way to capture a fortress was to cut off its communications and essentially starve out the garrison. This was never attempted along these northern waterways. Another way to attack one of these strongholds was to take it "by storm"—that is, launch an infantry attack against the fort's walls. Since forts are designed first and foremost with this type of attack in mind, it is an extremely costly method. The fort's cannons, the ditch and outworks, as well as its garrison firing from behind concealed positions would cost the attacker a significant number of casualties. More importantly, such measures were considered desperate and almost always resulted in a blow to the attacking army's morale as well.

The last method to seize a fort, a formal siege, was the most common. A formal siege came down to a duel of artillery. The attacking forces would select a portion of the fortification to assault. They would then dig siege trenches in a zigzag pattern toward the fortification's wall until at the desired distance, a parallel trench, or simply a parallel, was dug and firing platforms erected for the besieger's cannon. Typically, the first parallel was erected to provide covering fire and, thus, was too far away to actually breach the fort's wall. The siege trench was advanced and additional parallels deployed until the attacker was in a position to breach the defender's walls. When the fort's walls had been breached, the besieger could then push troops into the fortress and overwhelm the garrison. This last step was seldom executed. The general agreement between armies of the day was that if the fort's garrison did not surrender once its walls were breached, or even in some cases when the fort's walls were clearly at the point of being breached, then the whole garrison could be put to the sword once the stronghold fell.[3]

Along the Old Invasion Route of the Hudson and Champlain Valleys, these strong earthen forts posed a number of problems for the attacker. First, in order to secure the enemy fortifications, they would have to transport heavy cannon through the wilderness to the fort in question. With most siege guns weighing from one to three tons, this was not a trivial task on the colonial frontier, especially in an area devoid of roads. Second, the attacker would likely be forced into launching a formal siege of the enemy fortress. This meant that the attacker would not only have to bring a larger force to invest the position and ward off any enemy sallies while he erected his siege guns, but he would also have to bring enough supplies to maintain this army in the field for at least a month. There was also one other problem the attacker had to face: lack of expertise or a military engineer to conduct the siege. While this latter problem would vanish during the last French and

Introduction

Indian War, up until then it was a concern that typically worked in favor of the defenders.

Of course, the defenders would make every effort to disrupt the siege through the employment of their own artillery, as well as by the strength of their defenses and the use of the garrison. For the defenders, a siege was a question of time. The longer they could hold out, the greater the chances of relief. Thus, in most cases the strength of the fort was a function of how quickly a force could come to its aid to either lift the siege or, at the very least, extend it until the attacker abandoned the effort due to the season or logistical concerns.

One of the most important elements working in favor of the defenders was the health of the attacking army. In that day and age, contact between large numbers of people coupled with field conditions and poor camp practices frequently bred illness and occasionally epidemic. That, in turn, not only sapped the attacker's strength but his morale as well. Indeed, pestilence was the silent guardian of many fortifications, and the forts along the Old Invasion Route proved no exception to this rule. During General William Johnson's 1755 campaign against Fort St. Frederic, almost a third of the army was on the sick rolls, seriously calling into question Johnson's numerical advantage and the wisdom of advancing on the enemy fort. Another colonial campaign the following year under Massachusetts general John Winslow suffered far worse consequences. Poor camp discipline and practices led to an epidemic that at one point was killing fifteen men a day. "The camp was nastier than anything I could conceive," one British officer wrote after visiting Fort William Henry. "Their necessary houses, kitchens, graves, and places for slaughtering cattle, all mixed through the encampment."[4]

With the introduction of British regulars and their regimented camp routines, the mortality rate from sickness plummeted to the point that in 1759, General Jeffery Amherst reported his army in an exceptional state of health and attributed part of this to the introduction of spruce beer, which many at the time believed had medicinal benefits.

The transport problems placed before the attacker also favored the fort's defenders. In fact, it was one of the best defenses the fort possessed. As a case in point, between 1690 and 1760, the British regulars and American colonials would launch eleven expeditions against French forts along the waterway. Of these, eight campaigns failed to even reach their objective, and one was defeated in battle near its target. In each case, transportation, or lack thereof, was a fundamental cause of the expedition's failure. For instance, the logistics behind General James Abercromby's campaign against

Introduction

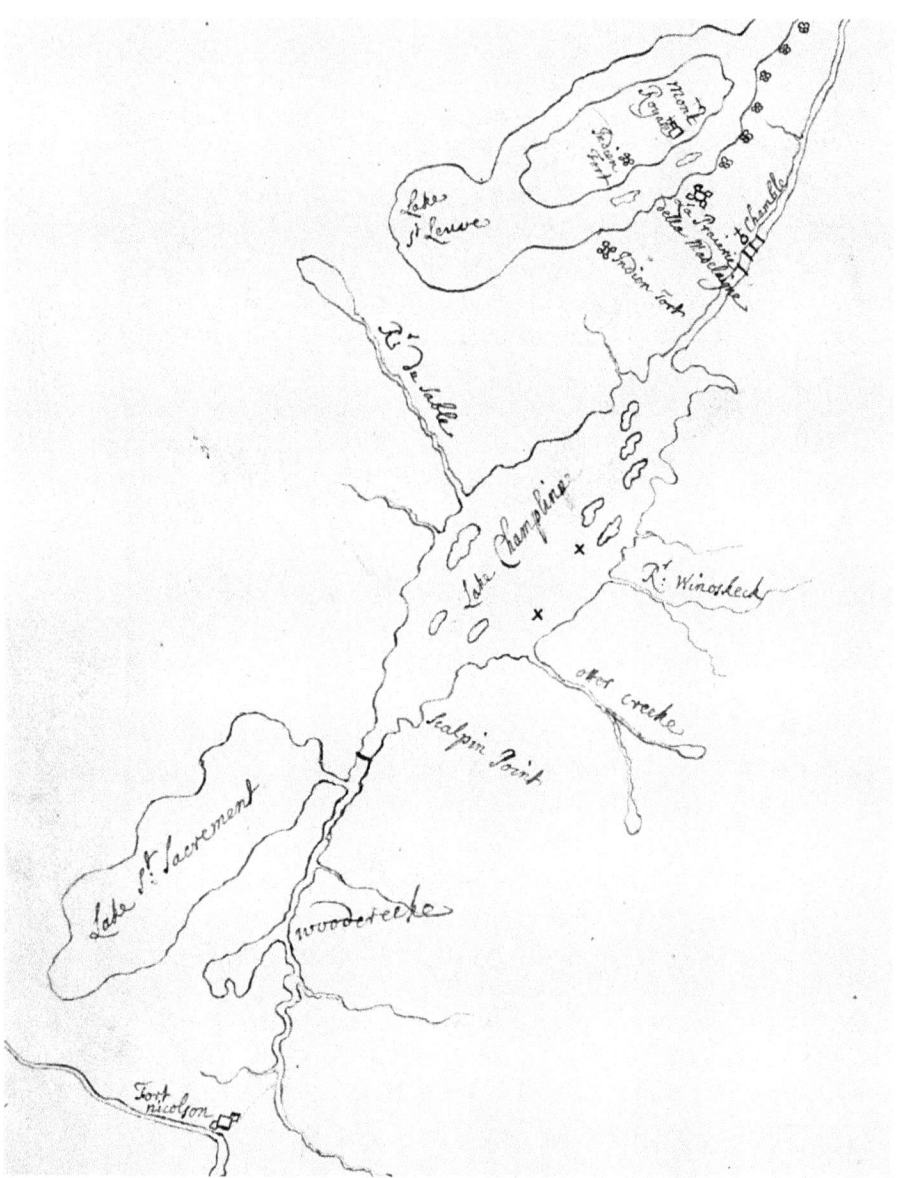

A circa 1709 map of the area between Fort Nicholson at the Great Carrying Place and Montreal. Note Scalpin Point or Crown Point, as it was known. At this time, the name was applied to both sides of the lake. Today, Crown Point is the name for the west side and Chimney Point the name for the east side. *From Hulbert,* The Crown Collection of Photographs of American Maps, *Series I/1 (1907).*

Introduction

Fort Carillon in 1758 were staggering. In order to support his army of nearly twenty thousand men at the head of Lake George, Quartermaster General John Bradstreet calculated that 1,500 bateaux would have to be constructed, a task he estimated would take until mid-May so long as the materials and skilled labor were available. The latter proved hard to come by as the New Hampshire carpenter company under Colonel Nathaniel Meserve, who had served in the area for the previous two campaigns, had been assigned to participate in the assault on Louisbourg. This created a skilled labor shortage, and it was not until June 7 that the last boats were completed. While this solved the water transportation problem, Bradstreet was still short close to five hundred men to operate this fleet, a situation that was only rectified when Abercromby assigned colonial troops to the quartermaster.[5]

The land transportation portion of the supply chain was even more daunting. When asked what would be required along these lines to support Abercromby's forces at Lake George and transport enough provisions to supply the army in the field for a month, Bradstreet estimated that eight hundred wagons and one thousand oxcarts would be required to accomplish the task. Even more depressing was that the estimate did not include the transportation needed to move the siege guns and their munitions to the lake or the time required to repair the roadways and bridges along the route.

Clearly, the army's requirements far surpassed anything to be found in the area. Bradstreet searched the three upper counties of New York with power to impress any transportation he came across, but in the end, he was only able to procure three hundred oxcarts and wagons. With such limited resources available, it wasn't until June that the necessary supplies and equipment were in place at Fort Edward. Even this required a Herculean effort on the part of the troops, who were often forced to drag cannons and supplies by hand over the muddy portage roads due to the lack of draft animals.[6]

The French, while they launched fewer assaults on enemy fortifications, also suffered from transportation issues. To keep Fort St. Frederic and Fort Carillon on Lake Champlain supplied required a fleet of small vessels and, more importantly, a major manpower commitment to operate these vessels. To cope with these problems, the French government finally decided to build several small schooners and sailing barges to ply the waters between these strongholds and Fort St. Jean to the north. While such an arrangement proved satisfactory during the spring, summer and fall, winter was another matter. Occasionally, expeditions did reinforce the forts during the winter months by carrying their supplies on sleds over the frozen lake, but these were not to be relied on. This, in turn, forced the garrison to stockpile

Introduction

enough supplies to carry them through the winter, which also made them vulnerable to a late winter or early spring campaign, when their supplies would be at the lowest point.

While a handful of strongholds was constructed early on, the last French and Indian War (1754–63) saw a substantial increase in the construction of fortifications along the waterways. After a failed campaign against Fort St. Frederic by four thousand American colonials in 1755, the French responded by building Fort Carillon on the Ticonderoga peninsula. From this latter vantage point a dozen miles south of Fort St. Frederic, the French could better control access to Lake Champlain by barring a landing at the outlet of Lake George as well as passage down the lake from South Bay or Wood Creek. The British and colonials revived the wooden forts along the upper Hudson and built Fort Miller to guard the Great Falls and Fort Edward to secure the Great Carrying Place. Farther north, at the headwaters of Lake George, colonial troops built Fort William Henry in the fall of 1755, which was to be employed as a staging point for future expeditions down Lake George against Forts Carillon and St. Frederic.

As it was, Fort William Henry would never host such an army. Seizing an opportunity while the majority of the British forces were involved in an expedition against Louisbourg on Cape Breton Island, the French army, under the leadership of the Marquis de Montcalm, ascended Lake George and after a brief siege captured the fort and its garrison in the summer of 1757.[7]

This would be the last major offensive by the French along the waterways. By 1758, the tide had shifted toward the British. With British naval dominance once again asserted in North America, New France was slowly being whittled down as fewer and fewer supplies and reinforcements reached the colony. While Montcalm would successfully defend Fort Carillon in 1758 from an army five times his size, such miraculous victories could hardly be counted on. The French commander aptly summarized the plight of the colony when he pointed out that New France could win a dozen battles and be no closer to victory, or it could lose one and lose everything.

As such, with the colony's resources dwindling, it was agreed to abandon both Fort Carillon and Fort St. Frederic as soon as the British appeared in force before them. The French army would retreat down the lake to Île aux Noix in the Richelieu River and from there make a last stand to protect Montreal.

For the British who put more men in the field every year, it simply became a question of time and securing their gains. Fort Carillon and Fort St. Frederic both fell to General Jeffery Amherst's army in the summer

INTRODUCTION

of 1759, and although Amherst was unable to advance any farther, he secured Lake Champlain by destroying the French fleet operating on these waters. He also managed to repair Fort Ticonderoga (Fort Carillon) and build a new fort at Crown Point that would be the launching pad for the following year's campaign.[8]

By mid-August 1760, the final acts of the French and Indian Wars on the Old Invasion Route were unfolding. Brigadier William Haviland, with 3,400 British regulars and colonials supported by an impressive artillery train of thirty-seven guns, set sail for Île aux Noix on August 11, 1760, and by the afternoon of August 23 had begun a sustained barrage on the French fortifications. For the defenders, it was a nearly impossible task to prevent the British from cutting the log boom that blocked passage down the Richelieu River. When what was left of the French fleet was captured below the fort a few days later, effectively cutting the defenders' communication to Fort St. Jean, the post was abandoned.

The end came quickly for the last French forts on the Richelieu River. Fort St. Jean was burned to prevent its capture by the advancing British, as was Fort Sorel at the mouth of the river. The last fort to fall was Fort Chambly a day before the end of the war. Strangely, here we see one of the last customs of a formal siege enacted. The fort's commander, Paul-Louis Lusignan, was formally requested to surrender. He refused, however, not because he believed the fort was capable of repelling the enemy but because honor demanded that he could not surrender his position without the enemy having even fired a shot. Colonel John Darby, in command of the British detachment, shrugged at the reply and ordered two mortars loaded. Just to make sure that Lusignan understood the peril of his decision, Darby formed a human wall of the town's citizens in front of his guns. He then gave the order to fire, and a pair of thumps echoed over the basin. Both rounds detonated within the confines of the structure, and with custom and honor now satisfied, Lusignan lowered his flag a few moments later.[9]

When one looks at the handful of forts built along the great north–south waterways, it is difficult to picture how effective these isolated posts became given their true strength. While almost every fortress drew criticism, in most cases the critiques failed to capture the compromises and difficult decisions that came with erecting the strongholds in the first place. The French, because of their limited resources, extracted huge benefits from these isolated posts, even if they were not perfected works or taxed the available supply system. The sheer number of campaigns launched against Fort Carillon and Fort St. Frederic and the fact that it took the British—who could annually put

Introduction

A two-part map of the Hudson and Champlain Valleys, circa 1759. This inland water route would be contested from the early days of the French and English colonies up through the War of 1812. *Boston Public Library, Norman B. Leventhal Map Center.*

Introduction

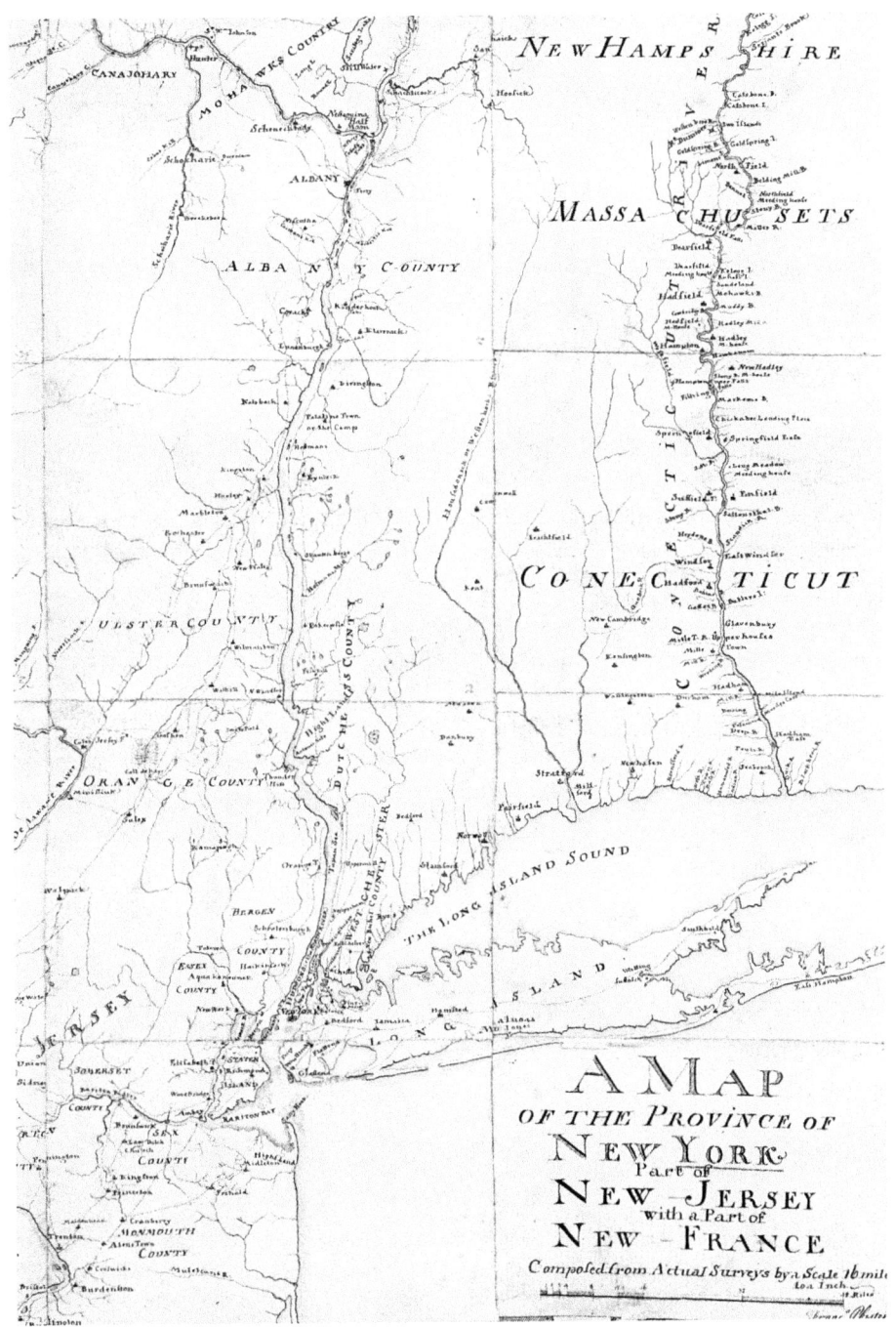

Introduction

tens of thousands of men into the field—four years to seize these outposts speaks to the importance of the structures. The Marquis de Montcalm, in speaking of Fort Carillon in particular and the chain of forts that supplied this post in general, simply referred to it as "the key to the continent." Few of his opponents would disagree with him. Sir William Johnson, who led an unsuccessful campaign in the area in 1755 and was present at the Battle of Ticonderoga in 1758, expressed a similar opinion, calling the location "a very dangerous and important place to be secured at all events, as it will then command the only two passes they have into our country."[10]

The need for the fortifications along the Hudson, Champlain and Richelieu waterways would not end with the surrender of New France, and many of the forts created during the colonial period would go on to participate in the American Revolution and, in some cases, even the War of 1812. New forts would be built as well and armies raised that at one point or another would besiege almost every post from one end of the waterway to the other. With the conclusion of the War of 1812, the armies withdrew, the fleets mothballed and the age of fortifications along the waterways entered its last chapter. The British would build Fort Lennox at Île aux Noix, and the Americans, after an aborted attempt, would build Fort Montgomery to the south, but neither fort would ever have a shot fired at it in anger. With relations between Great Britain and the United States steadily improving, the garrisons and resources committed to these positions slowly dwindled. Fort Lennox was abandoned in 1870, and with the closing of Fort Montgomery in 1926, the age of forts along the Old Invasion Route would officially come to an end.[11]

Chapter 1

THE RICHELIEU VALLEY

With the exception of Fort Amsterdam in New York, the Richelieu Valley forts are among the oldest on the Hudson-Champlain Corridor. The first of these, Fort Richelieu, was constructed near the outlet of the Richelieu River in 1643. Looking to stem Iroquois incursions into the St. Lawrence Valley, the governor of Quebec, Charles Hualt de Montmagny, selected a site at the mouth of the Richelieu River for a fort. Accompanied by one hundred soldiers and workmen, the governor arrived at the confluence of the Richelieu and St. Lawrence Rivers on August 13, 1643, and the next morning work on the fortification began. Trees were felled, stumps torn from the ground, ditches dug and the land about the site cleared of debris. The details of what was to become Fort Richelieu have not come down to us, but it is not hard to envision its general form. Given the resources available and the lack of military engineers in the party, the fort must have been nothing more than a crude wooden palisade with perhaps provisions for firing platforms along the walls, a watchtower or two at its corners and a ditch dug about its perimeter.

As it was, the fort was to be tested almost the moment it was completed. On August 20, the last wall of the fort was still being raised when a band of three hundred Iroquois, emboldened by their recent raiding successes, loosed a deafening war whoop and dashed on the structure from the nearby woods. Surprised, the French raced for their arms as the Iroquois charge carried them to the very walls of the fort, where they discharged their muskets into the compound through the firing ports cut in the walls. Seeing the Iroquois about to scale the ramparts, a corporal named Du Rocher gathered a score of men about him and launched a hasty counterattack that drove the Iroquois back to the edge of the woods. With the roar of gunfire echoing about him, Montmagny, who was on his barque when the attack began, made his way to shore and, with sword in hand, put his rattled troops in order just before the Iroquois rallied for a second charge. This time, the native warriors met stiffer resistance. Several of the leading braves fell victim to the coordinated defense, one later being found to have had his shield and body riddled with no fewer than a dozen musket balls. The Iroquois assault stalled and, under a hail of musketry, faltered. The two sides continued to snipe at each other for a period of time while a few of the more resolute Iroquois paddled toward the three barques anchored in the river, but the ship's cannon and a few well-aimed volleys quickly put an end to their plans. With no sign that their enemy was about to yield and their casualties mounting, the Iroquois broke off the attack and, as suddenly as they had appeared, retreated into the depths of the forest.

The French had one killed and four wounded in the engagement, including the governor's secretary, who had a musket ball lodged in his shoulder. As for the Iroquois, they seemed to have gotten the worse of the engagement. Several were killed, and probably twice as many were wounded. This aside, the victory was far more important than the casualties indicated. Surprised and outnumbered, the French had stood their ground, and although it had been touch-and-go for a time, they had defeated the warriors of the Five Nations. Given the long string of Iroquois victories over the last few years, such things could not be underestimated. News of the victory swept through the colony, and for all involved, it seemed to be a godsend, but in reality, it was nothing more than a brief reprieve.[12]

As for Fort Richelieu, it was finished and garrisoned without incident, but in the long run, it failed to accomplish its primary purpose. The fort did sever the water route used by the Iroquois to enter the St. Lawrence Valley, but Iroquois war parties responded by simply carrying their canoes around this obstacle. If the structure had been strong enough to accommodate a large garrison, routine patrols from the fort might have intercepted more war parties or, at the very least, forced the Iroquois to pull their canoes out of the water earlier and travel farther to remain undiscovered. But this was not the case. The colony could spare neither the manpower nor the resources to follow this course of action. Undersupplied and undermanned, Fort Richelieu became nothing more than an isolated stronghold in a vast wilderness, capable of defending itself but nothing more. Over the next

A portion of a 1688 map showing the Richelieu Valley forts. The first three forts were built in 1665. The following year, a fourth fort, named La Motte, was constructed on the northern end of Ile La Motte in Lake Champlain. The forts, built in a short period and under difficult circumstances, helped secure the Richelieu waterway, barring one of the favorite Iroquois invasion routes into the colony. *Library of Congress, Geography and Map Division.*

few years, the garrison of twenty or so and their Huron and Algonquin visitors found a safe haven so long as they stayed within the range of the fort's cannon. Outside this, however, just a few hundred yards away, the forest still belonged to the Iroquois, who lurked in the gloom, patiently waiting for an opportunity to pick off a wayward sentry or fall upon a wood-cutting party. By the fall of 1646, the colonial authorities, pressed for both men and funds and frustrated with the fort's inability to cope with the increasing number of Iroquois raids, ordered the fort's commandant, Jacques Rabelin, to abandon the post.[13]

The fort's supplies and a few of the smaller cannons were transported back to Quebec, the larger cannons that could not be removed were spiked and the structure was put to the torch. Rabelin's efforts, however, seem to have been halfhearted, for the following spring, the governor sent Jean

Bourdon and thirty men to ensure that the abandoned fort would be of no use to their enemies. When they arrived, they found that the Iroquois had finished what Rabelin had started, leaving the post a gutted shell. With little else to do, Bourdon and his men retrieved the spiked guns that had survived the flames and returned to Quebec.[14]

It would be another twenty years before a fort was erected along the shores of the Richelieu. In early 1665, after numerous pleas from government and clerical leaders of New France, King Louis XIV agreed to send a French infantry regiment to the colony to help deal with the Iroquois threat. When the elements of the Carignan-Salières Regiment arrived in Quebec over the summer of 1665, one of the first orders of business was to fortify the Richelieu Valley. This was done in part to hinder Iroquois war parties from using this route and in part as preparation for a military expedition against the Mohawk, the easternmost tribe of the Iroquois Confederacy.

The first of the planned forts was to be built at the base of the Richelieu rapids, about forty-five miles upriver and just above what would later become known as the waters of the Chambly Basin. With little more than the approximate location sketched out on a crude map, the task was handed over to Captain Jacques de Chambly and four of the recently arrived companies of the Carignan-Salières Regiment. On July 23, 1665, the party, accompanied by one hundred local workmen and a number of Indian guides, set out from Quebec in a flotilla of shallops and canoes.[15]

Chambly and his men ascended the Richelieu River without incident and selected a site for the fort along the west bank in early August. The fort was built in the same manner that Fort Richelieu had been a generation before. The soldiers and workmen quickly cleared the site of debris and dug a series of deep ditches that traced out the position of the fort's walls. Freshly cut 15-foot logs were then placed vertically in the ditches and lashed together, and the trenches were filled in with burnt earth and gravel to support the entire structure. In its final form, Fort St. Louis,, as the structure was christened, resembled a palisaded star measuring approximately 150 feet to a side with a protected main gate and redans protruding from the remaining three sides to allow the defenders to sweep the walls with musket fire. Along the interior of the walls was an elevated firing platform on which the troops could stand to fire through loopholes or firing ports cut into the log walls. Several crude wooden buildings within the fort's courtyard were the only structures and served as both storehouses and barracks for the garrison.[16]

While Chambly was busy raising Fort St. Louis, farther north Captain Pierre de Saurel and his company had been ordered to proceed to the site

The Richelieu Valley forts, circa 1665. From Jesuit Relations 49, 266.

of old Fort Richelieu and rebuild the fort. Saurel arrived at the mouth of the Richelieu in early September. Although he had fewer men at his disposal than Chambly, his task was much easier. The land about the burned fort was already cleared of any major obstacles, so only minor work needed to be done along these lines. The existing trenches, which held the former fort's walls, were excavated and new walls were raised in the same palisade fashion. When this was accomplished, a number of buildings were added to the interior to handle the garrison's supply and billeting needs. Finished, the new Fort Richelieu was a near copy of its former self. A pair of bastions were located at the corners of the easternmost wall and covered the landward approaches to the fort, while three redans were placed along the riverside or western wall of the structure. The customary interior firing platforms and loopholes were installed, and to break up any attack, a shorter picket wall of sharpened logs known as a *pieux liez* was added in the cleared area around the fort.[17]

The task of erecting the last of the three forts to be built that summer was given to the regiment's commander, Colonel Henri de Chastelard, the Marquis de Salières. Salières and his detachment of seven companies left Quebec on September 2 and by the twenty-eighth had reached the newly completed Fort St. Louis,, where they were welcomed by Captain Chambly and his detachment. With only vague orders on the positioning of the last fort and unfamiliar with the terrain that lay ahead, Salières spent the evening speaking with Chambly and his Canadian volunteers on the best spot to erect the structure. Fortunately for Salières, Chambly and some of the friendly natives had already reconnoitered the river ahead and identified a location for the third fort.

Having heard enough, the colonel and his troops departed Fort St. Louis the next morning, bundled as best they could against the slicing autumn rains. The detachment spent the day marching along the west bank of the river until they reached the site Chambly had recommended the night before, a small peninsula that extended out into the river, approximately eight miles upstream from Fort St. Louis. Here they made camp on the soggy grounds while the colonel consulted with his officers on the job ahead. Like Chambly and Saurel, Salières had no military engineer to consult with on the construction and layout of the fort. Not that it would have mattered much, for the materials and the number of tools available would have probably placed anything beyond what he had seen at Fort St. Louis out of the question. To complicate matters, the late arrival of the regiment's supply ship, the *Jardin de Hollande*, meant that they had no cooking pots or any means by which to prepare a warm meal, and by now the weather had deteriorated into a dreary blanket of chilling autumn mists and rain.

It was perhaps fortunate that Salières was present. Much of the colonel's well-known irritability stemmed from his decisive nature, and it was this spirit that propelled the project forward. Salières ordered the site to be cleared and, in keeping with his character, set the example by being one of the first to start on the task. The work progressed quickly enough that by October 2, much of the site had been cleared and the outline of the fort was laid out. The immediate trees, those cleared from the site and those in the nearby forest were cedars, unsuited for the walls of the fort because of their short length but perfect to cook by or to warm oneself against damp autumn air. The nearest stand of trees suitable for the fort's walls was located on a nearby island. Once again, the cedars were put to good use, this time in the form of several rafts, which were used to ferry men, equipment and felled trees between the two sites. As trees were stockpiled, work began on excavating the trenches for the palisade walls.[18]

With an insufficient number of tools, much of the grueling work had to be carried out by hand. Rain soaked through the troops' clothes and pooled within the trenches as they labored to expand the channel. Roots and stumps were torn out of the ground with branches used as levers or by sheer muscle

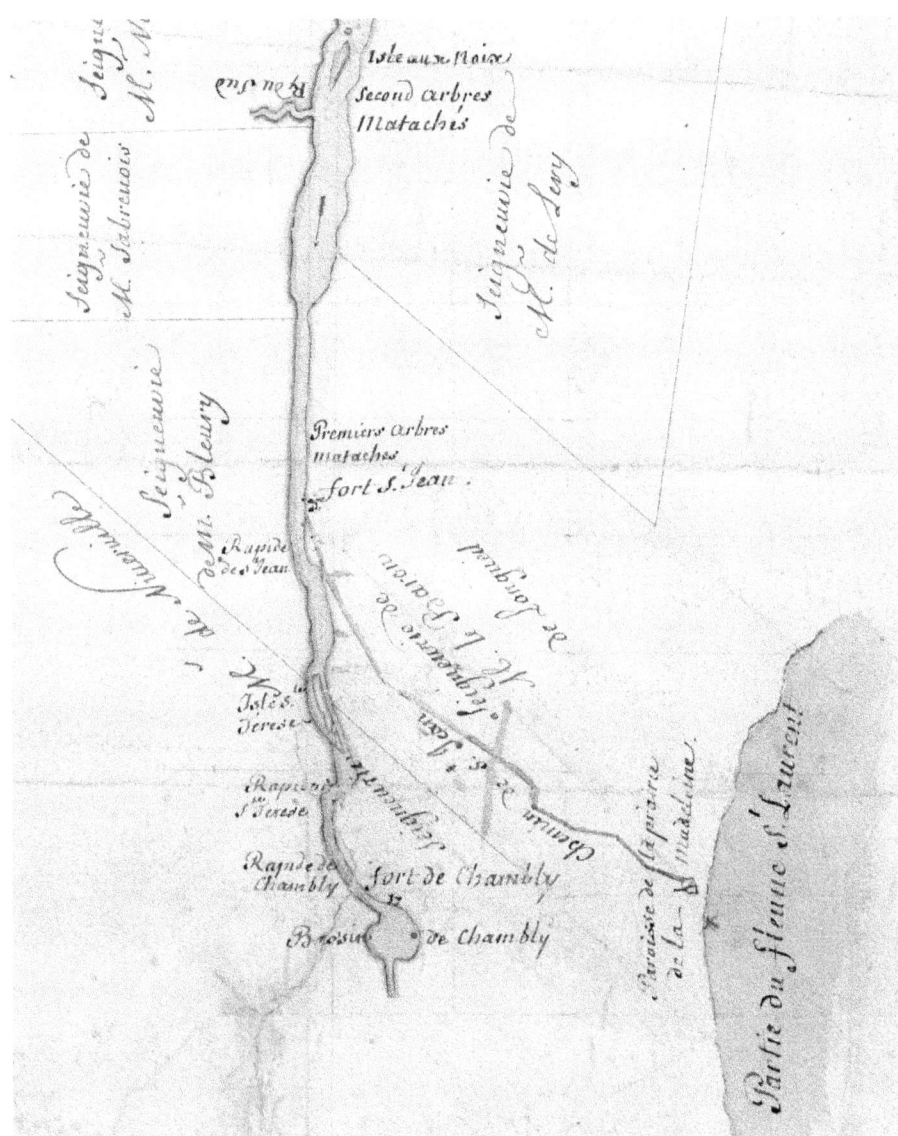

A circa 1750 French map looking south from Fort Chambly to Ile aux Noix, showing land grants and key positions along the Richelieu River. Note the rapids between Fort Chambly and Fort St. Jean. These would lead to the construction of the latter fort. *Boston Public Library, Norman B. Leventhal Map Center.*

when that didn't work. By October 12, the trenches were ready and the walls of the fort began to take shape. Within a few days, the outer walls had been completed and work shifted to constructing the courtyard buildings and interior portions of the fort.

Fort St. Therese, as it was called because it was finished near the day that honored the saint, was perhaps the most regular of the forts built along the Richelieu that summer. Its 12-foot walls were laid out in a rectangle approximately 130 feet long by 100 feet wide with a bastion occupying each corner. Its interior firing steps and the log buildings used to house the garrison and their supplies differed little from Fort St. Louis and Fort Richelieu, and like the other two, work would soon begin on a trench and a *cheval de frise* that would encircle the stronghold and present yet another obstacle to any attacker.[19]

The following year, a fourth fort, Fort St. Anne, was built on Ilse La Motte in Lake Champlain. The four bastioned wooden structure boasted fifteen-foot-tall curtain walls and, although smaller, resembled Fort St. Therese in form. The fort was abandoned half a dozen years later, having proven too difficult to supply, especially in the winter. For the next century, the ruins of this French outpost would become a common reference point on the lake for sailors and soldiers on both sides.

The Richelieu Valley forts improved Canada's defense of the waterway and aided in acting as staging points for a number of punitive expeditions against the Mohawk. As it would turn out, however, Fort La Motte was not the only fort to fall out of favor. As Fort St. Therese was positioned between two sets of rapids, its usefulness was soon questioned, and it, too, was abandoned not long after Fort St. Anne. Fort Sorel would continue a quiet existence until its destruction by retreating French forces in the closing days of the last French and Indian War.

Fort Chambly, however, was too important a position to be ignored. Located at the end of all southern navigation for heavy vessels entering the river via the St. Lawrence, Fort Chambly was a necessary outpost. In the winter of 1702, the thirty-seven-year-old fort was severely damaged in a fire. With relations between the two colonies deteriorating, there was little choice but to rebuild the fort. The second fort was a wooden structure like the first. The walls were fashioned from vertical logs supported by a filled ditch and a network of horizontal support logs. The old redans were replaced with bastions at the corners of the fort, and along the south wall a building called the "King's Store" was constructed, which functioned as a barracks and storehouse. A second building of a similar nature appears to have also been placed along the inside of the north wall, and the powder magazine was placed below the southeastern bastion. After the English colonies threatened invasion in 1709, it became clear that the current wooden fortification was insufficient. The next year, Governor Phillip Vaudreuil ordered Josue Boisberthelot de Beaucours to construct a regular stone fortification over the site of the current wooden one.

The fort was to be capable of withstanding only light artillery fire, primarily because it was considered unlikely that the English could haul heavy cannons over Lake Champlain. Beaucours laid out the new Fort Chambly in a square with three-story-tall bastions at each of its corners. Curtains, 30 feet high and constructed of limestone and masonry, linked the bastions, making the entire structure approximately 168 feet to a side. A fortified entrance was placed on the west wall facing the Chambly Basin.

The bastions and the wall along the side facing the Richelieu were pierced for cannons, and firing ports were set into the four-foot-thick walls. Long buildings were built against the east, south and west walls. These buildings served as the storehouses, workshops and barracks for the garrison. The length of these walls was also protected by a ditch, a shallow glacis and a covered way. Powder magazines and covered wells

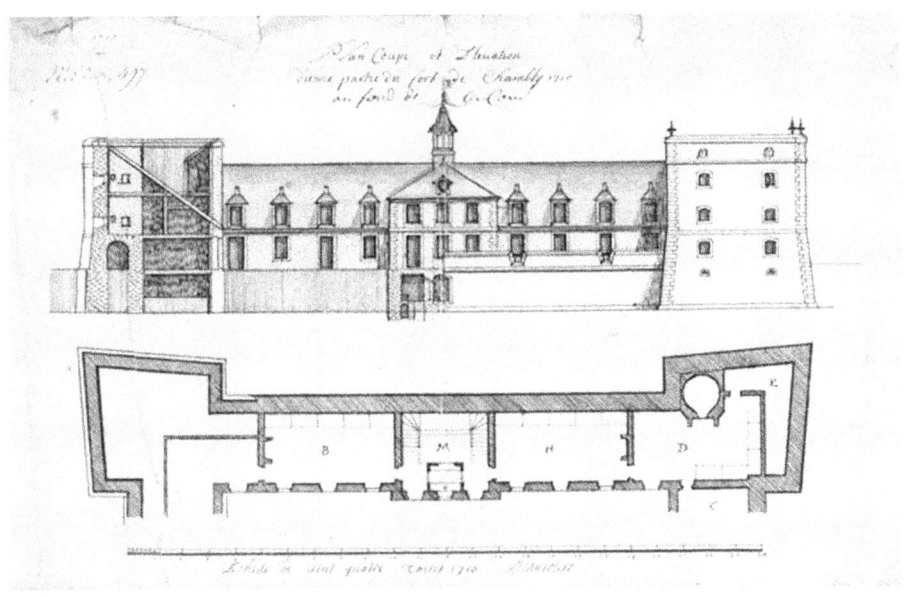

1710 plans for the reconstruction of Fort Chambly in stone. The project would take several years, but in the end, it succeeded in making Chambly one of the strongest posts on the frontier. *NAC, C-15990.*

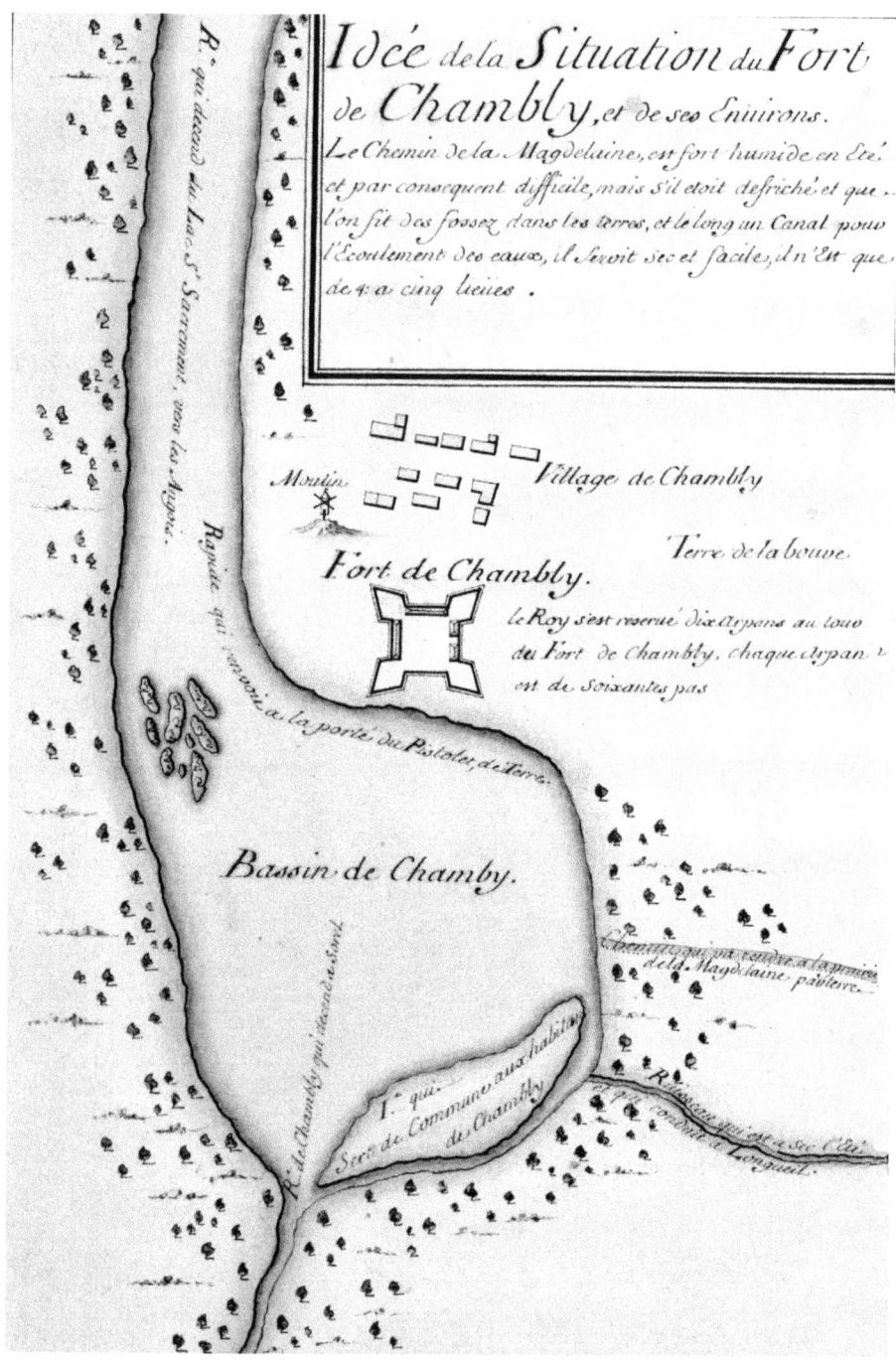

A 1717 plan of Fort Chambly and the small village that had grown up near the fort. Note that south is at the top of the map. *Boston Public Library, Norman B. Leventhal Map Center.*

within the lower parts of the bastions completed the works. When it was finished in the summer of 1712, it was an impressive structure, so much so that Governor Vaudreuil informed the minister of the marine that Fort Chambly was "now beyond insult."[20]

The fort underwent a major modification between the years 1718 and 1720. The colonial engineer Chaussegros de Lery rebuilt the north wall, which was already showing damage from the periodic ravages of the Richelieu River. De Lery also added watchtowers to the bastion corners; reworked some of the embrasures and loopholes; and oversaw the construction of a ditch, drawbridge and machicolation to better protect the fort's gate. In the 1730s, the fort again had its north wall reworked after years of flooding threatened to undermine its integrity. It was a persistent problem, and it would not be the last time that such repairs had to be undertaken. Fort Chambly was typically garrisoned by thirty to fifty troops of the Free Companies of the Marine. In addition, there were approximately forty families who lived in or about the hamlet that had taken shape around the fort. In a crisis, these individuals would take refuge within the fort, effectively doubling its garrison.[21]

While one of the strongest posts on the French frontier, Fort Chambly would ultimately be undermined by its location. With French expansion into the Champlain Valley via the construction of Fort St. Frederic at Crown Point in the late 1730s, Fort Chambly saw a boom in activity. At this point, all supplies headed south to Fort St. Frederic were first transported to Fort Chambly and then carried via wagon to the headwaters of the St. Jean rapids. Here they were loaded into small boats and transported south.

Fort St. Frederic had been in place only a few years when it became clear that the fort's supply chain was incapable of meeting its demands. Forwarding supplies from Chambly was a tedious and expensive proposition given the sets of intervening rapids and the limited capabilities of the small bateaux and canoes that traversed the route. What was needed was a heavier vessel, plying an established supply route. Such a vessel would serve another purpose as well. Armed with small cannons and superior to anything the English could slip past Fort St. Frederic, it would ensure French control of the waters. Soundings had pointed out that the lake was more than capable of taking such vessels, and although not all the navigation hazards had been charted, enough had been found that by the early 1740s, it was agreed to build a thirty-five-ton vessel on the lake. The problem was where to build it and where to anchor the vessel when it was not at Fort St. Frederic.

Governor Charles Beauharnois, intendent Gilles Hocquart and engineer Chassegros de Lery discussed the issue at length and arrived at a solution. A

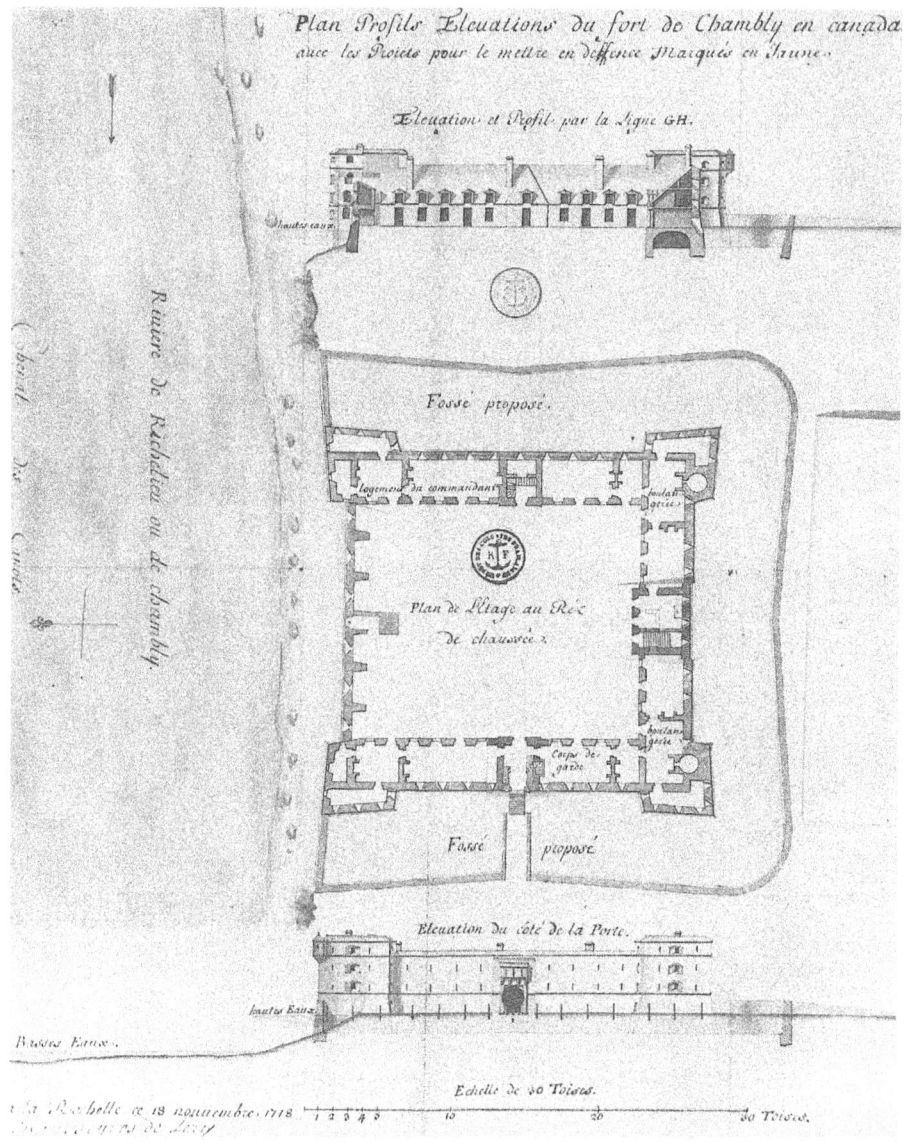

Fort Chambly as it stood in 1718. The original palisade works at this location were replaced with a stone fortification in 1710. Over the intervening years, a great deal of work was performed on this structure, in part to strengthen its defenses and in part to repair water damage from the ravages of the Richelieu River. *NAC, C-15886.*

wooden fort should be constructed on the west bank of the Richelieu River just south of the St. Jean rapids, where there was still sufficient water for a small schooner. The fort would act as a supply depot and magazine for Fort St. Frederic and a staging point for reinforcements going to its relief. The idea was attractive but suffered from one problem: since the water route to the proposed fort was impractical, a road would have to be cut to the fort from La Prairie, which, although an expensive proposition, was agreed upon as the only practical alternative.[22]

As appealing as the idea seemed to be, it was not until 1748 that work began on Fort St. Jean, and even then it was only undertaken because of the imminent threat posed to Fort St. Frederic during King George's War. De Lery's son, Gaspard-Joseph, oversaw the actual work, and his inexperience led to several setbacks. Although the land about Fort St. Jean was wet and sandy, Gaspard failed to sufficiently prepare the ground, which led to problems with the fort's chimneys and foundation settling. The cost overruns were enormous, leading the new intendent, François Bigot, to admit that he could not actually calculate the construction costs. His best estimate was that forty to fifty thousand livres had been spent, an incredible sum for a picket fort, especially considering that many of the materials for the project were cannibalized from the abandoned Fort St. Therese a few miles away.

Construction of the fifteen-mile road to La Prairie was fraught with similar problems. The land between the St. Jacques River and St. Jean consisted of long, marshy meadows and low-lying woods crisscrossed by small streams and dark clouds of mosquitoes and gnats. A total of 260 men toiled for three months in the oppressive heat, transporting and placing earth to fill in the road, but to little avail. After even the slightest of rains, the track turned into a sea of knee-deep mud. When the Swedish naturalist Peter Kalm traveled the road a year later, he referred to it as "unrivalled in wretchedness, wet and winding, so that my horse sank in the mire up to his belly in most places." On several occasions, a series of drainage ditches was proposed to manage the runoff, but such ditches were never constructed, and the road remained a source of irritation for years to come.[23]

When it was finally finished, Fort St. Jean differed little from the other picketed forts of the area. It had four bastions, with the two facing the Richelieu River being a story taller than the inland pair and constructed more in the fashion of blockhouses than open firing platforms. Firing parapets and loopholes were added to the walls, and a number of small cannon and swivel guns were mounted in the riverside bastions. A number of warehouses (both inside and outside the compound), barracks and a pier filled out the works. Although the fort was only intended to provide defense against small raiding parties or marauding bands when it was activated in late 1748, the French governor was quick to point out that it could reinforce Fort St. Frederic within forty-eight hours.[24]

In 1752, French royal engineer Louis Franquet, who held the post of director of fortifications at Louisbourg, toured Lake Champlain and the Richelieu River. After inspecting Fort St. Jean, Franquet recommended that the east bank of the river, a point-blank shot from the fort's cannon, be cleared to prevent an enemy from establishing itself there. As for the fort itself, Franquet wrote, "It cannot be denied that it is too strong a construction against musketry, and too weak against cannon, and as it is not possible to drive [cannon] there except at great expense, and with great difficulty, it is enough to propose only a few changes that can make it safe from a *coup de main*."[25]

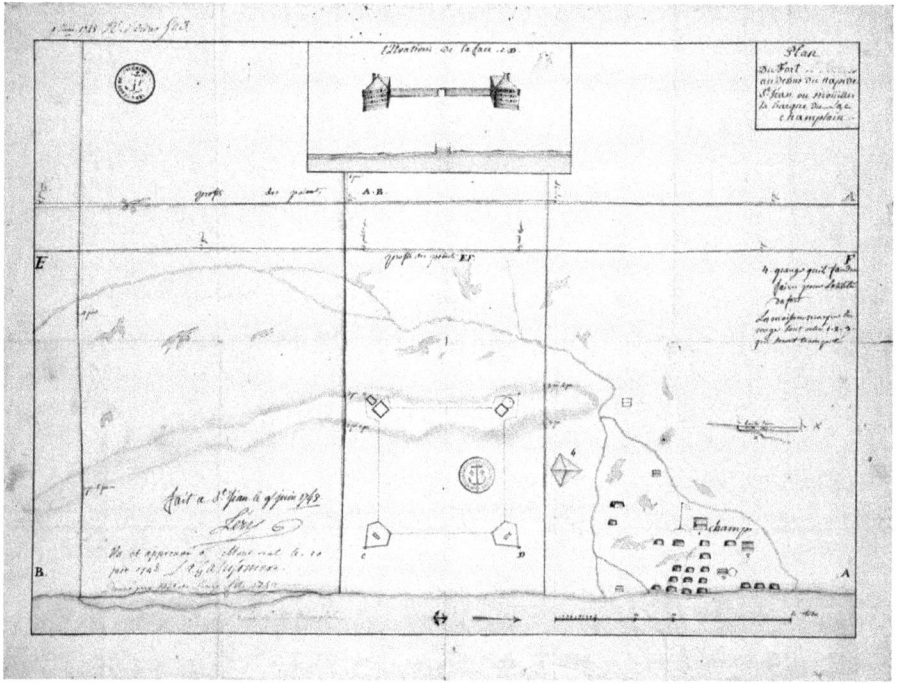

A plan of Fort St. Jean drawn by the fort's architect, Gaspard-Joseph de Lery, in 1748. Although a necessary post upon which Fort St. Frederic and later Fort Carillon depended for supplies, the construction of Fort St. Jean was mired with mistakes and cost overruns that eventually made it one of the most expensive palisade forts ever built in New France. Note that the Richelieu River, flowing from left to right, is located along the bottom of the map. *NAC, C-21780.*

Fort St. Jean would play a prominent role in French efforts in the last French and Indian War. The post was the primary supply depot for both Forts Carillon (Ticonderoga) and St. Frederic, which secured the French position in the Champlain Valley. It also became an important anchorage not only for supply vessels but also for a small fleet of warships built to contest control of the lake with the British.

The last fortified position built on the Richelieu during the colonial period was at Île aux Noix. Lacking resources and pressed on all sides by a more numerous opponent, the leadership of New France decided to abandon both Fort Carillon and Fort St. Frederic in early 1759. The army would withdraw to a fortified position on Île aux Noix near the head of the Richelieu River and, from there, contest any British move against Montreal. The fortifications at Île aux Noix were started in the summer of 1759, but they were not pursued in earnest until the arrival of the retreating French army in early August.

The island, named for its walnut trees, was located about twelve miles south of Fort St. Jean in the middle of the Richelieu River. A little over three-quarters of a mile long and a quarter of a mile wide, it divided the Richelieu into two channels. The western channel was the wider of the two, but the water here was too shallow most of the year, forcing heavier vessels to take the narrower eastern channel when passing the island. A group of small islets, often only a foot or so above the water, trailed the island, while upstream one could find another sandy islet barely above the waterline. Covered with walnut trees, maples and cedars, there was little beyond a patch of slightly rising ground on its southern end to distinguish this low-lying position from the surrounding countryside.

The position had both its pros and cons. First and foremost, cannons positioned on the island could dominate the river. This was especially true when cedar logs linked together with chains were erected as booms to block ship traffic through both channels. A bend in the river as one cleared Point a Margot moving north suddenly placed a ship one hundred yards from the southern tip of the island. This arrangement would at the very least place any English vessel attempting to bombard the island within the range of the defenders' cannon. The size of the island and its sandy soil also proved helpful. The sandy soil would limit natural shrapnel generated by the impact of enemy round shot and would tend to absorb the explosive power of their mortars' rounds. The sheer extent of the isle also meant that it was nearly impossible to cannonade the position into submission. An opponent could simply not bring enough artillery to bear to significantly degrade the island's defenses.

A portion of a 1763 map showing Île aux Noix and the outlet of the Grande Riviere Sud. Île aux Noix would prove to be one of the last French positions on the Old Invasion Route. *Library of Congress, Geography and Map Division.*

The extent of the island, however, was also a severe disadvantage in that the colony did not possess enough manpower and materials to sufficiently fortify the position. Nor were there enough men to garrison and defend the sprawling complex even if it was constructed. Beyond this, there were more serious drawbacks. Although the shoreline to the east and west was swampy and often flooded in the spring, as summer wore on, it tended to dry out. Nor was it by any means impenetrable to a determined enemy. As such, the island faced the very real possibility of being outflanked and cut off from its supply and communication lines to Fort St. Jean before the enemy even positioned a cannon.

The Riviere du Sud, which flowed into the Richelieu 900 yards downstream of Île aux Noix,, was also a constant source of concern. Should the English find a way to launch their vessels into this river, Île aux Noix would be isolated. Although General Charles Bourlamaque had been assured that the portage road to this river from the head of Missisquoi Bay could not handle heavy wagon traffic, this depended to a great degree on the time of year and the current rainfall totals. There was another point of concern as well. Although the island neatly secured the two narrow water channels leading north, it also placed the defenders at near point-blank range should the English manage to erect batteries on the opposite shores. The west shore was a little under 400 yards away, and the east shore was within extreme musket range a mere 250 yards away.

Bourlamaque, now in command of the island, was not pleased when he inspected the newly erected defenses. Although a great deal of work had been started by militia troops from Montreal, Bourlamaque found "the entrenchments badly done, without regularity" and "essentially defective." Nor were the entrenchments laid out to cover the entire southern approach to the island, forcing him to extend the works to both the left and right of their current position. At least a set of booms had been constructed that would block ship traffic along either channel. Bourlamaque concentrated his artillery on the southern end of the island and set his men to work perfecting the island's defenses. In all, it was difficult work over too large of an area for his force of just over three thousand regulars, marines and militia to contend with, but given time, he reported to Governor Vaudreuil, he could at least make the position impossible to take by storm.[26]

Fortunately for Bourlamaque, he would have the time. The British army under General Jeffery Amherst spent the summer waiting at Crown Point while warships were built to deal with the small French fleet operating on the lake. It was not until the afternoon of October 11, 1759, that Amherst sent his warships forward and loaded his men aboard their small boats. The weather halted Amherst's advance, but his warships did manage to corner the bulk of the French fleet at Crab Island, and within a few days, they were busy raising these vessels after the French scuttled their ships and made their way back to Montreal on foot.[27]

With the French fleet captured, the way to Île aux Noix was open to Amherst's army, but mid-October is a problematic time to be moving north along Lake Champlain—as the general's army would find out after several nights of rain and hard freezes. While the advancing season was a concern, so, too, was the result of General James Wolf's victory at Quebec the month

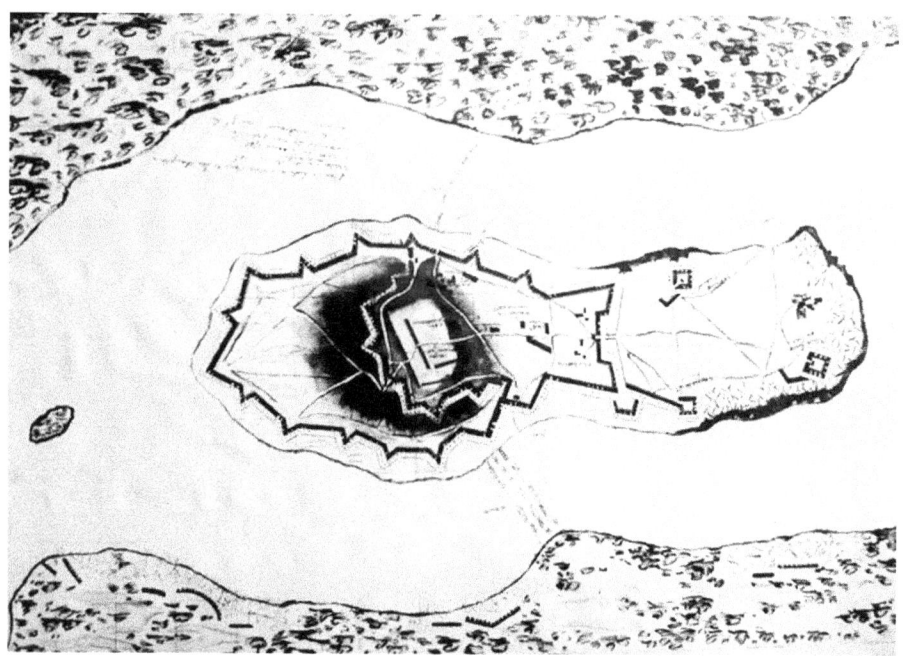

The fortifications of Île aux Noix as envisioned in 1760. In fact, the fortifications were not this advanced, particularly when it came to the northern horn-works, which covered the rear part of the island. *NAC, C-44000.*

before, which claimed the lives of both Montcalm and Wolfe. The defeated French army under General Gaston Levis had marched to Montreal, meaning that Amherst would now face the combined forces of Levis and Bourlamaque if he laid siege to the island. Taken together with the deteriorating weather and the naval superiority the British now had over the lake, Amherst wisely called off the campaign, and after garrisoning both Crown Point and Ticonderoga, he marched south into winter quarters at Albany.[28]

With all its advantages and defects, in the spring of 1760, command of this post was given to Montcalm's former chief of staff, Colonel Louis-Antoine Bougainville. Bougainville was forced to operate with less than half the men Bourlamaque had found necessary to defend the post the year before. Even with this setback, a good deal of progress was made. Working closely with colonial engineer Michel Lotbiniere, by August the island bristled with earthworks and field fortifications.

Most of the work was logically confined to the southern portion of the island, where the initial attack was certain to fall, and focused on improving the previously erected works. An eighteen-foot-wide ditch backed by an earthen rampart and parapet was laid out close to the water's edge, enclosing the southern end of the island in a zigzagging *U*-shaped line. The open portion of these entrenchments was then enclosed on the north by a pair of horn-works at each end and an interconnecting curtain wall. Behind the southernmost line of entrenchments, a second line of defensive works, laid out by Lotbiniere to oppose obvious English firing positions on the southeast shore, was constructed. This cut across the breadth of the island along the crest of the modest elevations found there. Enclosed within these fortifications on the southern section of the island were the various storehouses, barracks and the garrison's magazines, all interconnected by numerous footpaths that crisscrossed the open areas of the enclosure.

The northern part of the island, however, still lacked adequate defenses. Although the ground there was marshy, there was still the possibility that the English might attempt to bring cannons over to this part of the island and attack the French fortifications via trench work in the fashion of a formal siege. To diminish this possibility, both Bourlamaque and Bougainville agreed that a demilune should be constructed to protect the curtain wall between the two northern-facing horn-works. Given the time and manpower restrictions, however, this project was never undertaken. Instead, Bougainville had Lotbiniere erect two rectangular earthen redoubts near the center of the island about 225 yards from and roughly in line with the northern-facing horn-works. Such works were easier to put in place and had the benefit of not only covering the northern defenses but also disrupting any trenches the enemy might try to undertake.

The easternmost of these redoubts that was likely to come under artillery fire was eventually connected to the horn-work via a ditch and parapet arrangement, and it seems likely that, had time and resources permitted, the western redoubt would have had a similar arrangement put into place. Lastly, to dissuade the English from attempting a landing behind his southern defenses, Bougainville had a wooden blockhouse built on the far northeast corner of the island.[29]

The last element of Bougainville's defense was the vessels assembled around the island. The loss of the three xebecs the previous fall had left the French scrambling for naval support. The schooner *Vigilante* still remained, as did an armed sailing barge and the floating battery that Bourlamaque had constructed to guard the entrance to the east channel, but little else. To fill the void left by the loss of Lieutenant Jean d'Olabaratz's squadron, two row galleys known as tartans were constructed at St. Jean. The larger of the two, christened *Diable*, carried between forty and sixty oars and was armed with three eighteen-

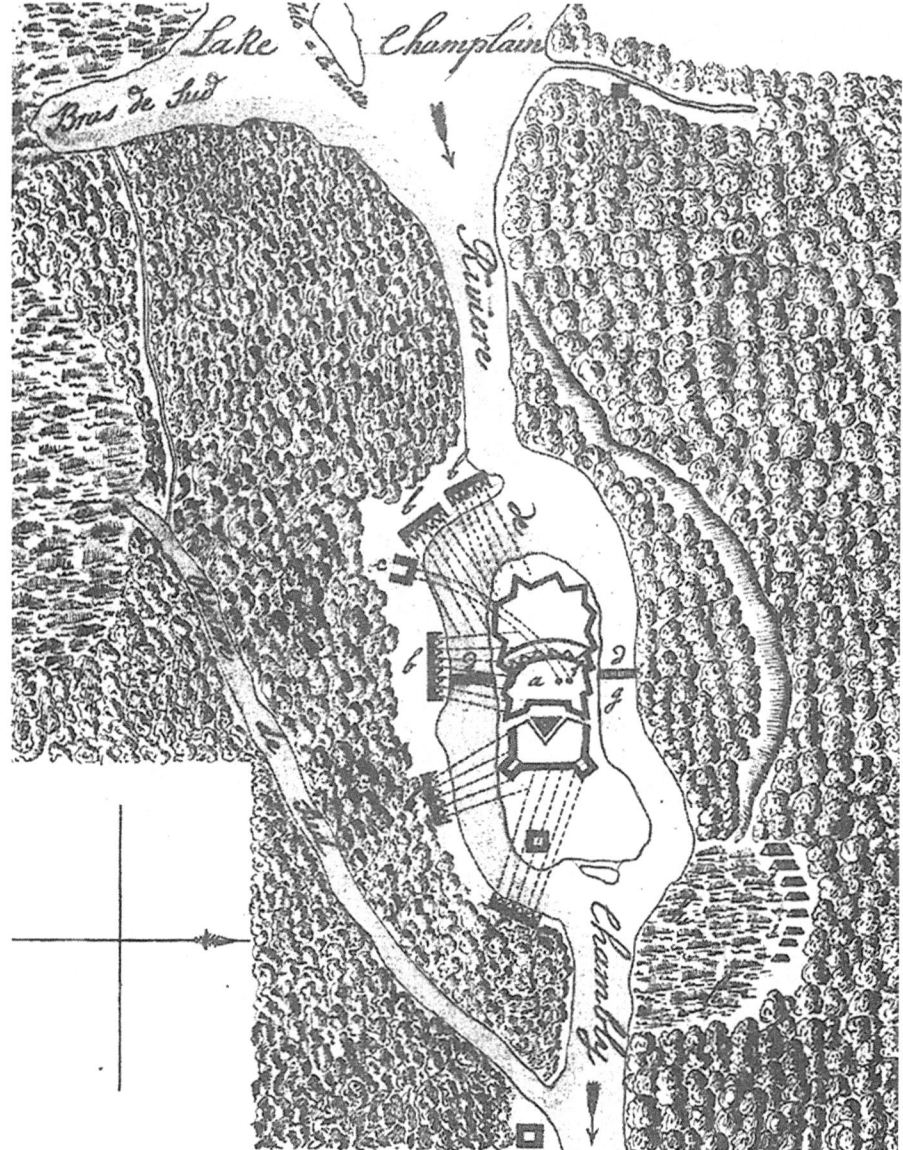

The siege of Île aux Noix, 1760, by the Chevalier Johnstone. *NAC, C-18292*.

Opposite: Louis-Antoine Bougainville, later admiral and count de Bougainville. This portrait shows Bougainville as he appeared during the American Revolution while in command of the French man-o'-war *Auguste* at the siege of Yorktown. Bougainville's appointment to command the fortifications at Île aux Noix left him under no illusions. At best, he could only hope to delay the English advance. *Library of Congress, Prints and Photographs Division.*

pound cannons. The smaller of the two, simply referred to as the "little one," carried twenty-four oars and was armed with a number of swivel guns and four-pounders. Four small gunboats known to the French as "Jacobs" rounded out the naval forces at Bougainville's disposal. These vessels carried a crew of a dozen or so, a few swivel guns and an eight-pound cannon typically firing forward.[30]

As it turned out, the fortifications of Île aux Noix were able to withstand a prolonged bombardment by a superior British force in late August 1760. As predicted, the island was simply too large to shell into submission, and successive attempts to cut the east boom resulted in failure. Instead, the fall of Île aux Noix can be attributed to the capture of Bougainville's little flotilla anchored below the island. Surprised by a detachment of Rogers's rangers and British infantry under Lieutenant Colonel John Darby, the French fleet was captured intact, effectively severing Bougainville's communications with Fort St. Jean and, more importantly, providing the British with a means to bypass the fortifications.

Bougainville abandoned the fort the next evening and marched north to Fort St. Jean. There, the army halted briefly to gather together what supplies could be carried off before setting the fort ablaze and marching northwest toward Montreal.

The last of the French Richelieu forts to fall was Fort Chambly. Tasked with quelling French resistance in the Richelieu Valley, Colonel Darby and Major Rogers appeared before the fort on September 4, 1760, with one thousand men and six cannons. As was the custom of the day, Darby sent a formal summons to the fort's commander, Paul-Louis Lusignan, to surrender the post. And as was custom, Lusignan refused, as no commander surrendered his fort without a shot being fired. The fact that Fort Chambly was not in a position to resist did not weigh into the commander's decision. Honor was at stake and demanded his response.

Darby shrugged at the answer and ordered a pair of his mortars to open fire. After a few shots—the only artillery rounds ever fired at the fort in its ninety-five-year history—honor had been served, and Lusignan ordered the British flag to be flown indicating his surrender.[31]

Chapter 2

LAKE CHAMPLAIN

A secure Fort Chambly was viewed as a necessity, but in the late 1720s, talk began to circulate about locating a fort farther south at Pointe-à-la-Chevelure, the region known today as Crown Point and Chimney Point. The idea was not new. Governor Jacques-Rene de Brisay, the Marquis de Denonville, had pointed out the need for stationing an advanced force at this strategic spot where the lake narrows to the width of about four hundred yards as early as 1686, but nothing had come of the suggestion. This time, however, Jean-Louis de la Corne, a distinguished officer in the troops of the marine, made a more striking case for fortifying the location. In the summer of 1730, la Corne, who had been to Pointe-à-la-Chevelure on several occasions, outlined his case in a letter to the governor of New France, Charles Beauharnois.

First, in practical terms, Fort Chambly was poorly located. A set of rapids existed between the fort and Lake Champlain. Canoes and light vessels could navigate from the fort onto the lake during high water, but larger vessels were restricted to either end of the rapids. It was clear that the best way to prevent an English invasion was to maintain control over the lake, a task that couldn't be accomplished from Chambly. Second, a fort farther south would force the English out of their boats sooner to deal with the outpost. This, in turn, would give the defenders of Canada more time to organize a response. Given that a well-equipped army, with a little luck in the form of good weather, could sail from Wood Creek to Chambly in a matter of a few days, any sort of delay imposed on their progress was crucial. Third, a fort farther south could be used as a staging and rallying point for raids into the upper Hudson Valley and would also place the Mohawk villages and New England within easy striking distance of the French, a consideration that would facilitate Canada's philosophy of *La Petite Guerre*. La Corne warned the governor that although peace currently existed between the two colonies, the English were not asleep. If action was not taken soon and the English allowed to establish themselves at this location, then "we could never show ourselves upon Lake Champlain except with open force, nor make war against them except with a large army."[32]

Beauharnois was in complete agreement and sent la Corne's memorandum to the king requesting permission to fortify Pointe-à-la-Chevelure. In the meantime, as it would take some time for a response from France, he dispatched a small party to Pointe-à-la-Chevelure to trace out and prepare the ground for a stockade fort.

Beauharnois's request met with a receptive court in Versailles. The governor was directed to take the steps necessary to erect a palisade fort at the location and ensure its defense until a more permanent structure could be undertaken. With the English established at Oswego on Lake Ontario, the French court did not need many reasons for agreeing to the idea, but as it turned out, they had another purely political reason for wanting a fort at Pointe-à-la-Chevelure. As the royal commissioners met to determine the limits of the English and French possessions in North America, the king's councilors at Versailles began formulating a "watershed" theory to define the colonial boundaries. In its basic form, it claimed that all the waters drained by the St. Lawrence River and the lands that were contained therein belonged to New France. In the case of Lake Champlain, it was true that the French held the right of discovery, but occupation was just as important when it came to making a boundary claim. Thus, the establishment of a post on Lake Champlain would not only bar an English army attempting to move north but would also solidify French territorial claims over the entire Champlain Valley. There was also another point that intrigued the court's imagination. The fort at Pointe-à-la-Chevelure could be used as a bargaining chip in the current boundary negotiations, perhaps even in an exchange for the English fort at Oswego.[33]

Work on the wooden fort began in August 1731 under the watchful eye of Rocbert de la Morandière, an officer in the troops of the marine. The task was nothing new for Morandière, who had built a dozen palisade forts across New France. The fort was laid out as a four-bastioned one-hundred-foot-by-one-hundred-foot structure located on a bluff near the water's edge

COLONIAL FORTS OF THE CHAMPLAIN AND HUDSON VALLEYS

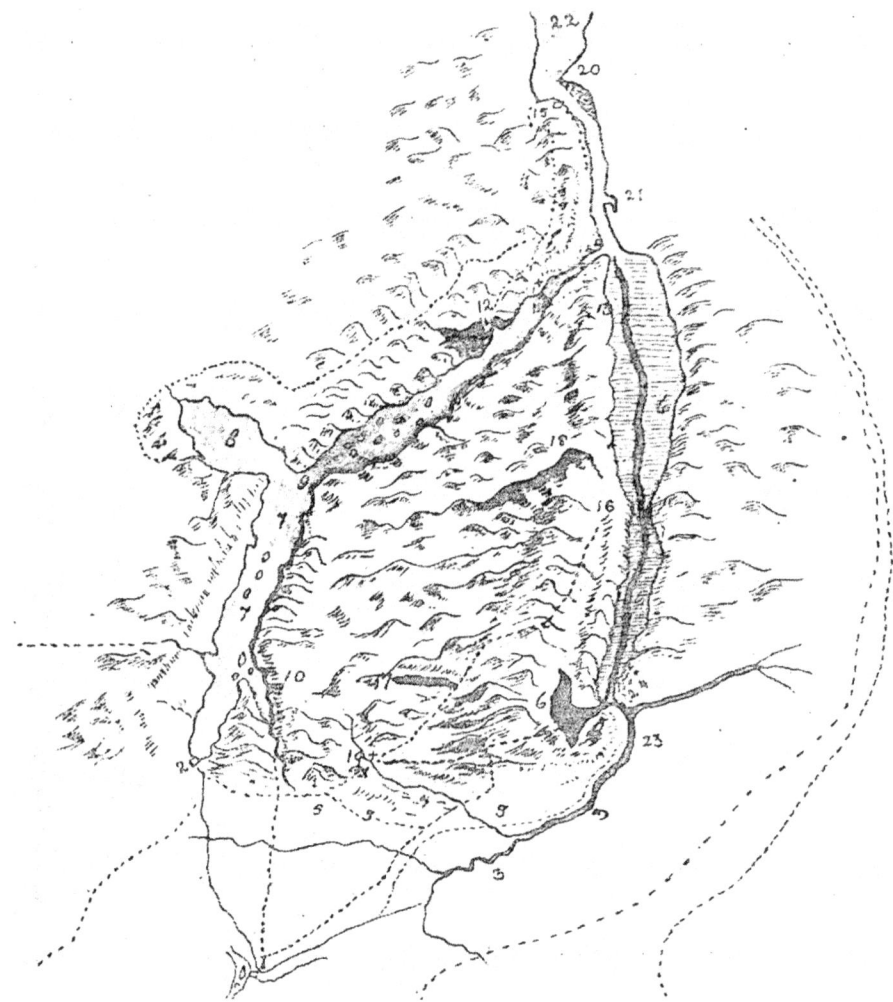

Sir William Johnson's 1755 sketch showing the Indian trails and features of the area between Fort St. Frederic and Fort Edward. *From Johnson Papers, II.*

Legend:

1: Fort Edward
2: Fort William-Henry
3: Wood Creek
4: Small Creek
5: Trail
6: South Bay
7: Lake George
8: Kanhusker Bay
9: First Narrows
10: Sakundawide Bay
11: Second Narrows
12: Remarkable Mountain
13: The Sugar Bush
14: Ticonderoga
15: Fort St. Frederic
16: Two Rocks
17: Small Lake
18: Remarkable Mountain
19: English advanced post
20: Crown Point (Chimney Point)
21: Presqu'Isle
22: Lake Champlain
23: Wood Creek Falls

on what is today called Chimney Point. Inside were a series of buildings for lodging the twenty-man garrison, their munitions and supplies. A chapel had even been constructed to meet the occupants' spiritual needs. By mid-October, the task was complete. So, too, was the letter of complaint from the English ambassador at Versailles, the Earl of Waldegrave, who had received numerous reports over the course of the summer on the French endeavor. The ambassador claimed that the fort was built on Iroquois lands and, as such, was in violation of Article Fifteen of the 1713 Treaty of Utrecht, which made the Iroquois British subjects. He called for the fort's immediate destruction and the withdrawal of any French troops encroaching on these lands. Contrary to issuing such an order, the French king responded by commending Beauharnois on his vigilance and ordering work to proceed on erecting a more permanent fortification at the site.[34]

In keeping with the king's wishes, Beauharnois asked de Lery to draw up a set of plans for a stone fort at Pointe-à-la-Chevelure. De Lery, who was responsible for the design and construction of most of the major fortifications in Canada, took to the task, and by the end of the month, he had completed a preliminary set of plans for a stone fort on the west bank of the lake at modern-day Crown Point. The governor winced when he examined the engineer's drawings. Instead of a regular bastioned fort, de Lery had centered his design about a four-story-tall *redoute à machicoulis*, a structure that resembled a rook or corner battlement on a medieval castle. He circled this structure with a more conventional stone wall in which the redoubt roughly occupied the position of the easternmost or water-side bastion. A ditch and a wooden palisade protected the outer wall, and the main gate, which was a small blockhouse in and of itself, was located along the northwest wall.

Taken as a whole, the design looked more the product of the Middle Ages than the military dictums of Vauban. De Lery must have noticed with some amusement the quick glances that passed between the governor and the intendent, Hocquart, as they pored over his drawings. "Unusual," one imagines the governor to have commented. "And expensive," the intendent would have returned. The engineer gave them their moment and then explained his work. Contrary to discarding Vauban, he had abided by one of his first principles—that being that fortifications should be adapted to peculiarities of the surrounding terrain. He had dismissed the idea of a regular bastioned fortress for several reasons. First, the long and harsh winters would make servicing and manning the cannons along the outer walls next to impossible. This was not the case with the current

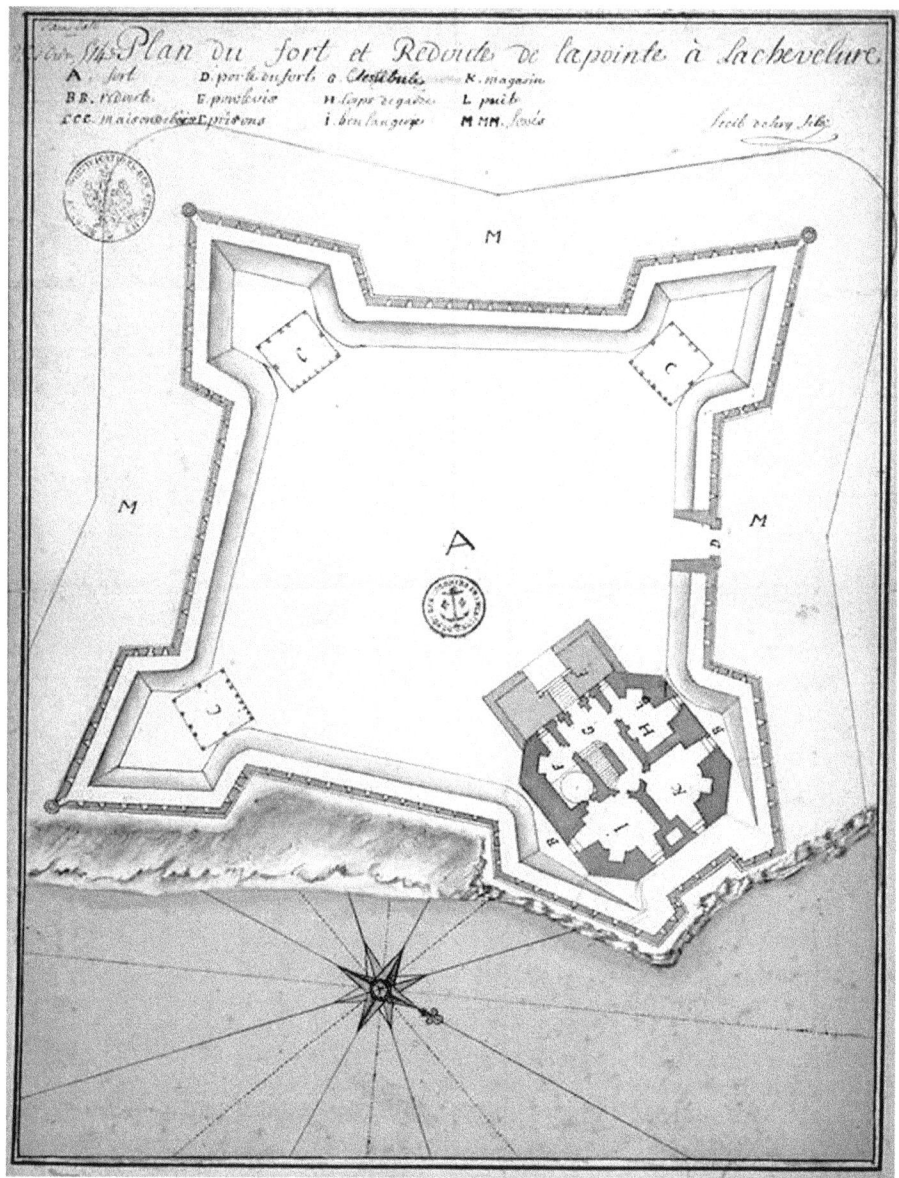

Plan of Fort St. Frederic, circa 1737. *NAC, C-21784.*

design. The fort's cannons were mounted in firing ports along the upper levels of the redoubt, all of which could be closed to the elements by the means of wooden shutters. Second, the redoubt would be difficult if not impossible for attackers to scale. The firing ports could be closed, and the other windows were covered with iron bars to prevent access. This was not true of a regular fort, and as such, a regular fort required a larger garrison to maintain its security. With his design, a few sentries posted at advantageous points could survey the entire landscape, which, when coupled with the natural security of the redoubt, meant a far smaller garrison, perhaps some fifty men, would be required.

There were nods of approval from the two men. The reduced garrison costs pleased Hocquart as much as the reduction in manpower pleased the governor. The outer walls, de Lery continued, like the redoubt, were loopholed and provided with a covered way that traversed their perimeter. When called upon, small cannons and swivel guns could be mounted along these walls to provide for a first line of defense, but like the keep of a castle, even if these walls were breached the defenders could still fall back on the redoubt, which overlooked the entire compound.

Beauharnois and Hocquart were convinced and transmitted the plans along with their recommendations to Versailles. De Lery followed with a letter to the minister of the marine outlining his thoughts on the matter. The court was at first receptive, but in April 1733, the minister of the marine expressed concerns over the costs involved and questioned whether the current wooden fort or a small regular stone fort wouldn't meet their requirements. No, the governor answered. No, the intendent seconded. The fort at Pointe-à-la-Chevelure was the first line of defense against the English. A palisade fort was fine for protection from raiders or marauding bands of Indians but would not last a day before English cannons. A more significant structure was required, and it was required soon, as there were already reports filtering in of English designs on the area. A statement in the form of a major fortification was required not only for the security of New France but to prevent further English expansion into the region. The court relented on the projected costs and gave the order to start work on what would be Fort St. Frederic early the following year.[35]

Additional delays meant that the actual construction did not begin until the spring of 1735. The task was anything but routine. Building a wooden fort was one thing, but constructing a stone fortification in the heart of the wilderness, particularly one with a four-story-tall tower, was quite another. Numerous tasks had to be completed before the work could even begin. Tools, provisions and beasts of burden all had to be transported to the site from Chambly and Montreal. A sawmill was set up not only to cut the timbers and planks needed for the fort but to fashion the construction ladders, platforms and hoists as well. A limestone quarry was established

Chief engineer of New France Gaspard-Joseph Chaussegros de Lery. Lery, who had served in France during the War of Spanish Succession, would be active in designing and constructing defensive works throughout Canada for forty years. *Collection du Musée National des Beaux-Arts du Québec.*

about half a mile away. From there, teams of oxen conveyed an endless stream of stones to the site, where they were cut by hand and fashioned into the blocks that would make up the fort's walls. At the same time, the foundation for the new fort was excavated out of the rocky peninsula with pick and spade. Throughout the summer months, crews of men toiled at the task, filling the tranquil landscape with the sounds of their effort. In late October, de Lery suspended work until the following spring. Illness among the men and the sheer effort of the task had put him behind schedule, but not as much as might have been expected. The garrison was currently lodged within the redoubt, and the works, although "not perfected," were for the moment sufficient to withstand an attack.[36]

The finishing touches were put on the fort nearly three years later. When finally completed, it was an imposing structure, a black limestone barricade rising out of the wilderness—a lock firmly affixed to the waters of Lake Champlain. The outer walls were twenty feet high and two feet thick with bastions protruding like the points of a compass. A twenty-foot-wide ditch crowned with a row of sharpened stakes circled the walls, creating a deadly trap for those who would foolishly elect to storm its ramparts. Within the flag-stoned compound sat a small stone church and numerous other buildings to house the garrison and their supplies. But it was the citadel that held the eye, a dark sentinel standing over the edge of the lake. From a distance, it might be mistaken for a lighthouse, but as one approached closer, he would discern the cannon ports and arched bomb-proof roof indicative of its true purpose.[37]

The fort was not without its critics. A stone windmill was placed just east of the fort and doubled as a lookout post, having a better view up the lake (south) than the fort. Peter Kalm, the Swedish naturalist who traveled the length of the lake a decade later, questioned whether the fort shouldn't have been placed at this location. French royal engineer Louis Franquet, who visited the fort a few years later, was far more critical. Franquet agreed with Kalm and pointed to the dominating ridgeline two hundred yards from the bay, or west bastion. The engineer suggested that a redan be built to help cover this side of the fort, but in the same breath, he pointed out that it would do nothing to prevent the enemy from employing his mortars against the fort from behind the ridge.

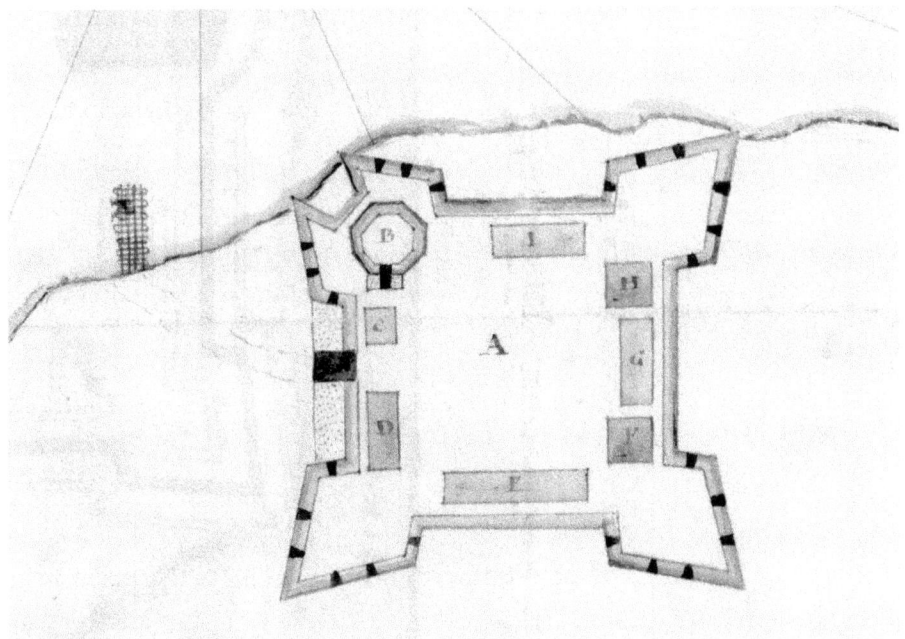

A circa 1755 plan of Fort St. Frederic. *Boston Public Library, Norman B. Leventhal Map Center.*

Legend:

A: Fort St. Frederic
B: The citadel
C: Officers' guard room
D: Garrison's guard room
E to I: Storehouses
K: Wharf.

Placement issues aside, Franquet was far more critical about the construction of the stronghold itself. The drawbridges were broken, several of the bastions required repairs, the soldiers' quarters were in ruins, the powder magazine door was weak, the chapel was too small, the tower leaked and needed shingles to prevent its early decay and even the latrines were falling down. However, one of the most pressing concerns, the engineer quickly noted, was water:

> *There is no water in the fort, and supposing it blocked, it would be necessary to open the door to go to the river* [lake], *a disadvantage which would lead to too many accidents. To remedy this defect, it is necessary to establish a cistern which will be easily filled with rain water, by means of the gutters to be placed around the cover of the redoubt.*[38]

The French engineer was also surprised the find the stone-faced walls already cracked in numerous places. "Moreover," he wrote, "most walls of the enclosure suffer; one would be inclined to believe that it is more by a defect of construction than by the influence of the land."[39]

In the end, as he had at Fort St. Jean, Franquet concluded that Fort St. Frederic was "too weak against artillery and too resistant against musketry." Even so, Franquet admitted that it would take a battery of cannon and a formal siege to secure its works—an act that would require a tremendous logistical effort on the part of the enemy, and one that would take a great deal of time. Time enough for an army to march to the fort's relief.[40]

During King George's War (1744–48), Fort St. Frederic served as a staging point for French and Indian raids into New York and New England. In response to this, several expeditions were organized by the American colonies to capture the French stronghold, but none of any consequence materialized. The fear of the fort was not to diminish in New York and New England, and shortly after the start of the last French and Indian War (1754–63), a major expedition under the command of General William Johnson was launched in 1755 to neutralize the stronghold. Johnson's army would never get farther than the headwaters of Lake George. After a large detachment of colonial troops was routed in a devastating ambush, the French commander followed up with an attack on Johnson's encampment. Johnson's men stood their ground and held off the French attack until after losing their commander, when the enemy broke off the engagement and retreated back to Ticonderoga. While Johnson technically won the battle by avoiding annihilation and being in possession of the battlefield at the end of the day, it had proven costly. Over three hundred colonial casualties littered the area, rattling the amateur soldiers. Just as importantly, it became clear to their commander that their will had been broken, effectively ending any serious threat to Fort St. Frederic by Johnson's troops.

It did not take long for the governor of New France, the Marquis de Vaudreuil, to realize that a fort at Ticonderoga was a necessity. When news reached him of the French defeat at Lake George, he dispatched colonial engineer, Lieutenant Michel Lotbiniere, to the Ticonderoga peninsula with orders for him to fortify this position as well as the portage at the outlet of Lake George a few miles away.[41]

A week later, Lotbiniere and several senior officers spent the afternoon examining the Ticonderoga peninsula under the watchful eye of a small escort. The terrain before them posed several problems. First, except for a rocky ledge near the center of the peninsula, it was covered with a dense growth of oaks, maples and birch trees, obstructing the party's view and making it nearly impossible to take accurate measurements. "I was obliged to operate in the midst of a wood," Lotbiniere later wrote to the minister of war, "without being able to see, while surveying more than thirty yards ahead of me." The second problem was even more difficult: exactly where to place the fort? The site was far from ideal. Across the lake about seven hundred yards away sat what would become known as Mount Independence, while a little over a mile to the southwest loomed the imposing figure of Mount Defiance, or Rattlesnake Mountain, as it was called at the time.[42]

Both of these heights overlooked the peninsula and were within easy cannon shot. Ideally, a network of forts would be called for that took all three positions into account, but Lotbiniere had neither the resources nor time to justify such an approach. No, the fort would have to be placed on the Ticonderoga peninsula, but exactly where was the question. The layout of the peninsula did not help matters. It rose quickly, almost cliff-like, from the water's edge along its south and southeastern borders. It then formed a brief plateau or elevated spine, which soon descended through broken, wooded terrain to its eastern and northern water's edge. These features were suitable and even desirable for the positioning of a fort, but along the west or landside, the ground slowly elevated to a ridge a few hundred yards away that completely dominated the site. Standard practice called for the fort to be placed here, at the highest local elevation, but doing so would undermine the purpose of the fort, which was to control the entrance

COLONIAL FORTS OF THE CHAMPLAIN AND HUDSON VALLEYS

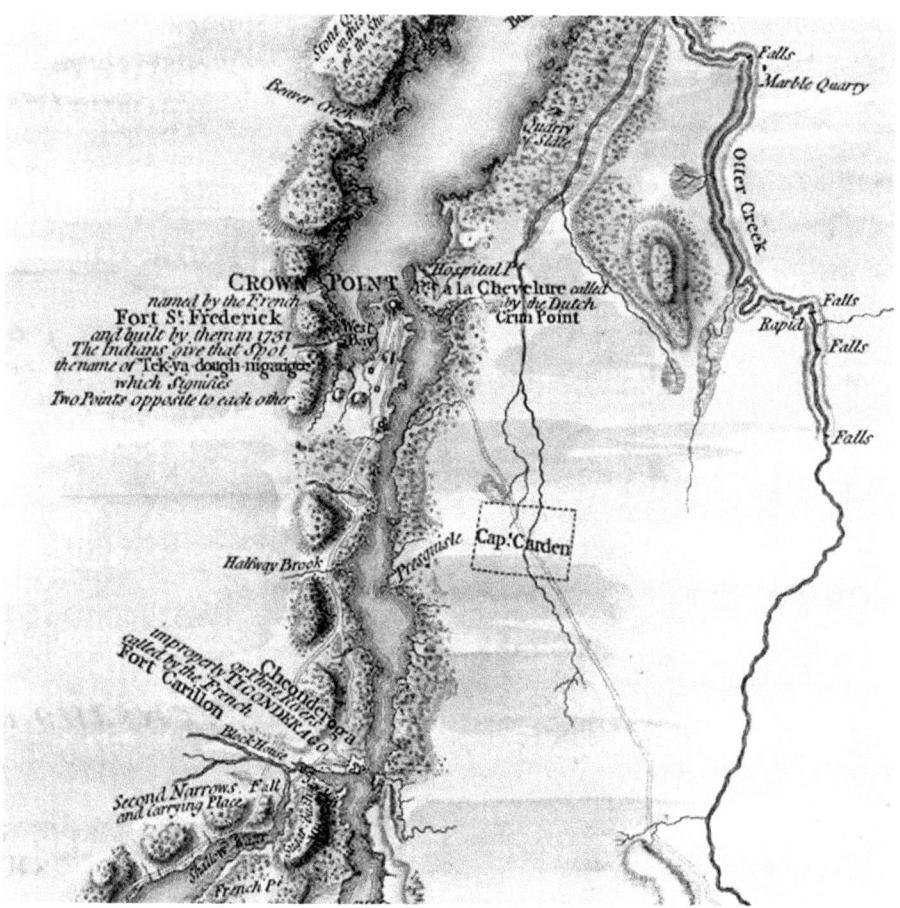

A portion of a 1762 map showing the area between Fort Carillon (Ticonderoga) and Fort St. Frederic at Crown Point. *Library of Congress Prints and Photographs Division.*

into and out of Lake Champlain. There was, of course, an answer to this problem, which as a trained military engineer Lotbiniere must have considered. The fort could be placed on the heights and a second fort, or fortified post, placed near the water's edge. Within support of each other, the pair of strongholds would accomplish both aims and prove stronger than any single fort, no matter where the other was positioned. But again, Lotbiniere was forced to compromise. The governor had made it clear that whatever fortifications were decided on had to go up that fall, and it would simply take too long to construct a pair of posts. Thus, with some reluctance, the rocky plateau near the water's edge was chosen as the most suitable site.[43]

On October 14, 1755, the French army, bolstered by several hundred recently arrived workmen and detachments from Fort St. Frederic, swarmed over the site. For days, the broken cadence of axes and the rasp of saws echoed across the lake. By the seventeenth, the ground had been cleared enough for Lotbiniere to trace out the fort and set the men to work digging out its foundation. He then turned his attention to the construction of a fortified encampment just below the fort near the water's edge. In actuality, the encampment was more pressing at this stage, for it was to serve the needs of the two thousand men working on the project and harbor both the provisions and supplies needed to erect the fort. Later, of course, it would attend to the needs of a visiting army, without whose support the fort would be unable to survive a conventional siege.

Overall, it was another monumental undertaking. Like Fort St. Frederic a generation before, everything had to be brought to the site. Tools, provisions, men, horses—all had to be carried by water from Montreal to the peninsula, a journal of a few days to a few weeks, depending on the weather. The first structures to go up were engineering sheds within the village taking form below the fort. These housed the tools and supplies. A small church, a makeshift infirmary and a temporary barracks or two were erected at this stage. The latter, however, seem to have been initially used for other purposes, for the men remained encamped in tents, which led to a steady stream of complaints as the morning frosts and wet winds of autumn set in.

By early November, the troops had hacked a fifteen-foot-wide and five-foot-deep outline of the fort from the rock and earth of the peninsula. It was an impressive task, accomplished for the most part by pick, muscle and sheer will. Lotbiniere had ruled out a stone structure, at least for this year. Instead, the recently felled oaks were put to good use. Fort Vaudreuil, as it was dubbed for the moment, was constructed just like its English counterparts to the south. It was a bastioned square, perhaps of a more symmetrical layout than Fort William Henry, but nonetheless encompassing the same general principles and form. Double rows of timbers were placed ten feet apart within the outline of the fort and secured to one another at intervals with dovetailed cross members. This hollow box was then filled in with earth from the trench at its foundation. By late November, the walls had reached seven feet, and a few crude buildings had been added within the compound to serve as barracks and storehouses for the garrison. A number of cannons sent by Vaudreuil had been brought down from Fort St. Fredric and were temporarily mounted along the ramparts and near the fort's gate.

To support these works and cover the boat landing, a fortified post, known as Lotbiniere's redoubt, had been raised along the water's edge. At this point, it was nothing more than a rude palisade of interlocking logs, but backed by a few small cannons and an alert garrison, it more than served its purpose. This was only the first of several outworks envisioned by Lotbiniere. A palisade about the village was on the list, as was a series of fortified camps along the La Chute River to guard the outlet of Lake George, but time had limited his efforts in the latter area to a few flying camps that were abandoned when the army departed for Montreal on November 28.

The architect of Ticonderoga remained until February in hopes of continuing the work with the fort's garrison, but apathy and the weather limited the effort to the completion of the fort's barracks. Realizing that nothing further was to be accomplished, Lotbiniere left for Montreal in February, disheartened and with the long list of tasks to be completed but confident that the fort was capable of defending itself.[44]

Lotbiniere returned to Fort Vaudreuil in early May 1756. He briefly toured the site to check on the state of the works undertaken during his absence and, after pronouncing himself satisfied, launched into the next phase of the fort's construction.[45]

The first order of business was the completion of the fort's bastions. Each had to be bomb-proofed, or casemated, to protect the garrison and their supplies during a siege. Of these, the southeastern, or Joannes bastion, as it was called, was the most pressing, having been designated as the fort's powder magazine. Unfortunately, the assignment required hacking a storage cellar out of solid rock. For weeks, the sound of pick on rock filled the campsite. When this was finished, two rows of heavy timbers were dragged into place to form the magazine's ceiling, and the intervening space between the two was filled in with dirt. Amazingly, it took less than a month to accomplish the task. By June 7, seven thousand pounds of powder and 4,500 rounds of ammunition were secure within the structure. "I can certify that it is one of the finest and best in all of Canada," one of the participants in the venture proudly recorded in his journal.

By mid-July, the bastion above the magazine had risen to a height of eight feet and was faced with stone blocks quarried a short distance away. In fact, almost all of the work from this point on was done in stone. The discovery and opening of the nearby quarry was, of course, one reason. But stone was hardly lacking on the Ticonderoga peninsula. More important was Lotbiniere's recruitment during his absence of a number of masons to fashion the material into a useable form. The two landside bastions, known

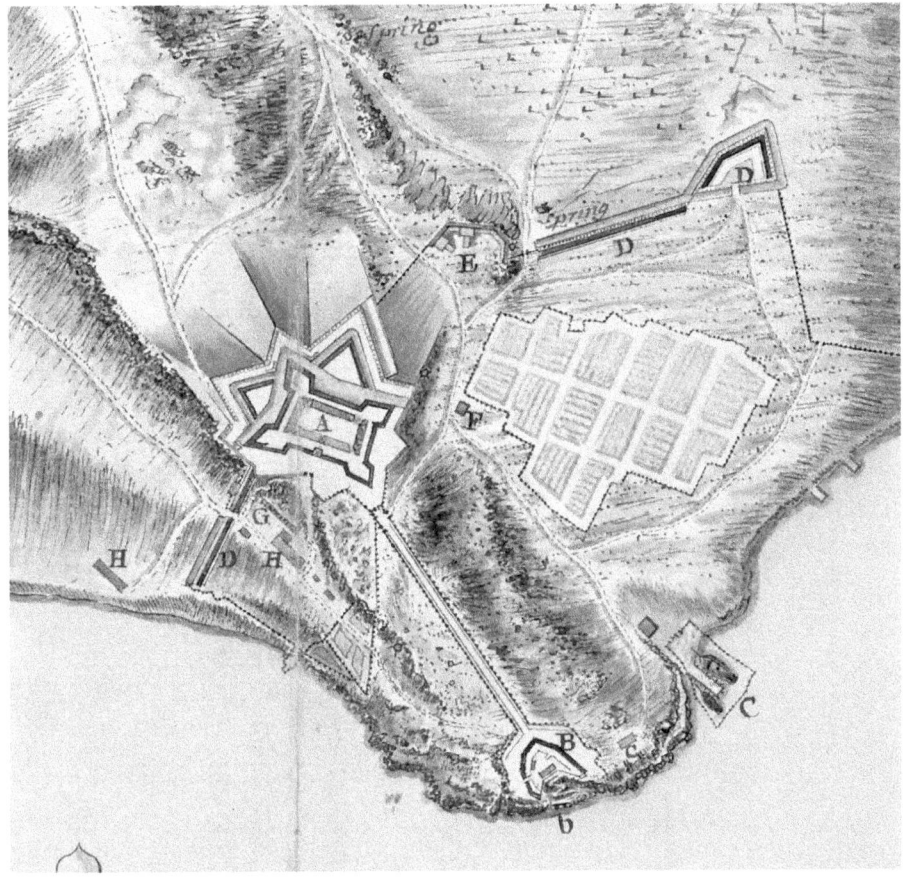

A portion of William Brasier's map of Ticonderoga, 1759. *Boston Public Library, Norman B. Leventhal Map Center.*

Legend:

A: Fort Carillon (Ticonderoga)
B: Lotbiniere's redoubt (b) Lower battery
C: Wharf and Harbor
D: Earthworks
E: Stoneworks
F: Well
G: Ovens
H: Storehouses

as La Reine and Germain, were the most likely to come under fire during an attack and, thus, were dealt with next. Here, too, chambers were hacked out of the rock foundation and casemates constructed. By mid-July, both were almost thirteen feet high, faced in stone and mounted a variety of cannons. By the end of August, the parapets and platforms had been added, completing the structures. The Languedoc bastion, which was the least exposed to an attack, lagged behind the other three, but given that it held the

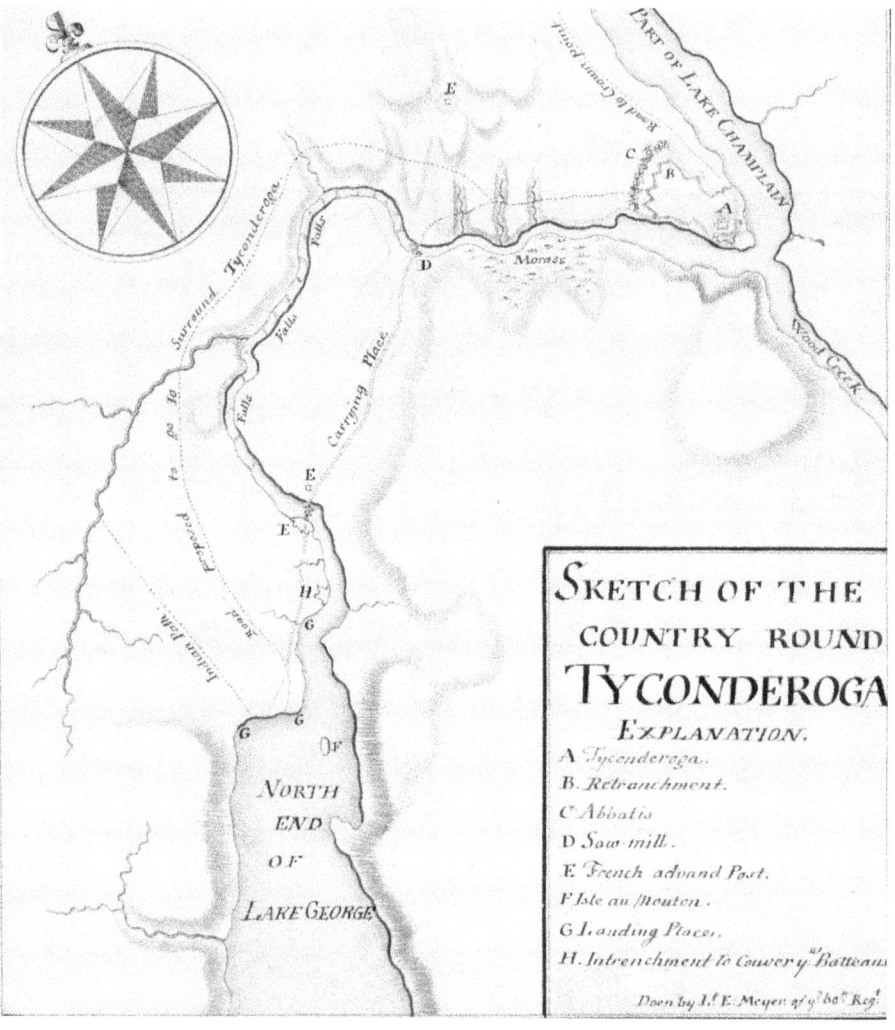

A 1759 map of the area and outlying fortifications about Fort Carillon. *Boston Public Library, Norman B. Leventhal Map Center.*

fort's bakery in its cellar, it could not be completely ignored. By late August, it, too, was well underway.[46]

There were a multitude of other matters to contend with as well. The redoubt at the boat landing, or Lotbiniere's redoubt, as it was called, was improved, and a hospital was added within its confines to isolate the sick from the rest of the garrison and expedite their evacuation should it become necessary. Soundings of the lake from Fort St. Frederic to Ticonderoga had been made the previous year, revealing a channel deep enough to handle heavy vessels. Thus, fashioning a rudimentary pier was one of the first orders of business, and on July 1, the schooner that normally plied the waters between Fort St. Jean and Fort St. Frederic made its first call at Fort Carillon. Although it may have passed with little notice to most, Lotbiniere was quick to inform Vaudreuil that the fort could now be resupplied within forty-eight hours at a significant savings to the king. Work also progressed on a "great shed," located adjacent to the fort, which overlooked the growing village along the southern shore of the peninsula. Although it might have been termed a shed and even used as such, the structure was really a large blockhouse fashioned of oak timbers laid one upon another and boasted several cannons on its roof when it was completed. Below this, the hamlet of Ticonderoga was taking shape. Warehouses and barracks were erected, a canteen was established and the small chapel raised the year before was now reworked in stone. Scattered through these structures were kilns, ovens and later a forge—all the necessary elements of a garrison town. By the end of the year, over a dozen structures dotted the shoreline, all enclosed within a sturdy wooden palisade.[47]

Although he had set the main works into motion, Lotbiniere spent a good part of his time occupied with the fort's outworks. Of these, the most important were the camps along the La Chute River and at the portage near the outlet of Lake George. The previous year, a pair of temporary encampments had been erected in this area, which, after being abandoned at the onset of winter, had been reestablished with the spring thaws.

Currently nothing more than a cluster of tents and a felled tree or two, a more permanent arrangement was envisioned for the area. In mid-May, Lotbiniere, Lusignan and Lieutenant Colonel Privat of the recently arrived La Reine Regiment conducted a survey of the area. One of the first items the trio agreed on was the location and establishment of a sawmill on the La Chute River. Although Lotbiniere was not explicitly authorized to construct one, a mill had become a near necessity, and the swift waters offered too much of an opportunity for the young engineer to overlook. The site, just below a series of shallow waterfalls, was also well suited for a guard post, and it was decided to permanently establish a sergeant's guard of fifty men at the location. Satisfied, the three men moved south along the portage trail behind a screen of grenadiers, who beat the surrounding woods for any signs of lurking Iroquois or English Rangers. About a mile away, they selected three sites for the advanced posts. The first two occupied opposite ends of a pontoon bridge that spanned the narrow rapids near the head of the river. The last, and farthest south, was at the outlet of Lake George where

Michel Chartier de Lotbiniere. A colonial engineer, meaning that he belonged to the French Department of the Marine, Lotbiniere was heavily involved in the fortifications of the Champlain and Richelieu Valleys during the last French and Indian War. Lotbiniere's work at Ticonderoga was severely criticized by Montcalm's French army engineers, but a good deal of this criticism was politically motivated as part of the running feud between Marquis de Montcalm and Governor Vaudreuil. *NAC, C-011236.*

the bateaux and canoes were typically pulled ashore. Given that there was already a flying camp at this location, all that was required was the addition of a redoubt and some basic fortifications to secure the position.

This was only the first of several surveys Lotbiniere was to conduct of the area. After some consideration, the fortifications erected at the portage were demolished and the site was shifted slightly south, which better covered the exit of the lake and provided the defenders with an escape route through a gorge that ran back toward the main encampment. In late June, the engineer returned to finalize his plans for the sawmill. A dam site was selected, and work was started on the building, the flume and the overshot wheel needed to harness the river. As a bridge was deemed necessary to link the mill to the peninsula, this task was undertaken as well. Lotbiniere then wrote the governor, somewhat after the fact, for permission to erect the mill and for the remaining items needed to finish the work.[48]

Earlier, Lotbiniere had ordered the ground cleared along the northwestern side of the fort, but as the work parties approached the ridgeline, he ordered a halt. Conventional practice called for the ridge to be cleared and an advanced post in the form of a strong redoubt constructed along this dominating ground to deny its use to the enemy and allow the fort's garrison an unobstructed view of movements along the heights. But as Lotbiniere paced the stump-strewn ground on a misty June morning, he saw something else. Hands clasped behind his back, he paused to gaze at the elevated tree line before returning to his pacing. A few moments later, he was back to his previous pose. Surely a worker or two noticed the lone figure, but by now they probably knew better than to give it more than a passing thought. After some time, the engineer glanced at the wooded heights one more time and nodded to himself. When he came back into the camp, he gave the order that the trees along the heights of Carillon were not to be disturbed. No doubt there was a puzzled look or two, probably followed by a shrug or a slow, perplexed nod. As it turned out, he had guessed correctly that the fate of the fort might one day rest with this belt of trees.[49]

With the outlying projects completed or well underway, Lotbiniere turned his attention back to the fort. A pair of stone barracks capable of holding four to five hundred men was started within the fort. Lotbiniere's insistence on completing these seems to have put him at odds with several of his superiors, who felt their priority was misplaced, but in the end a compromise was reached, and Lotbiniere, who had spent the better part of the last winter at the fort, pushed the task forward so that both were completed before the first snows. During this time, the outer sections of the fort were started. Two demilunes were laid out to protect the landside curtain walls, and the last of the ditch was blasted out of the rock about the perimeter of these fortifications and along the length of the adjoining walls. Preliminary work was also done on the glacis, and a counterguard was traced out for the La Reine bastion, but none of this was anywhere near complete by the close of the year. The pace of the work and the tasks involved nearly overwhelmed the engineer. "I have been so much occupied by details," he wrote to the minister of war, "pushed to minuteness here, where, despite of himself, a man must work at all trades, possessing no means of relief, that I have found it impossible for me to execute my observations."[50]

Both Montcalm and Levis arrived amid this commotion on July 3. The next day, the two officers began a systematic assessment of the position. Neither was happy with what he found. Montcalm informed Lotbiniere that the works were progressing too slowly. The engineer agreed, but the issue had initially been one of manpower, not purpose.

Levis and several thousand troops would remain to finish the works while Montcalm led the rest of the army in an attack on the British forts at Oswego. It proved to be an uneasy summer for the new commander of the newly christened Fort Carillon. Not only was he busy dealing with excursions by Robert Rogers and his rangers and watching a colonial and British army encamped at the head of Lake George, but even the elements seemed poised against him when a lightning strike started a fire in some of the ongoing works. Levis was correct in that the joint colonial and British army force

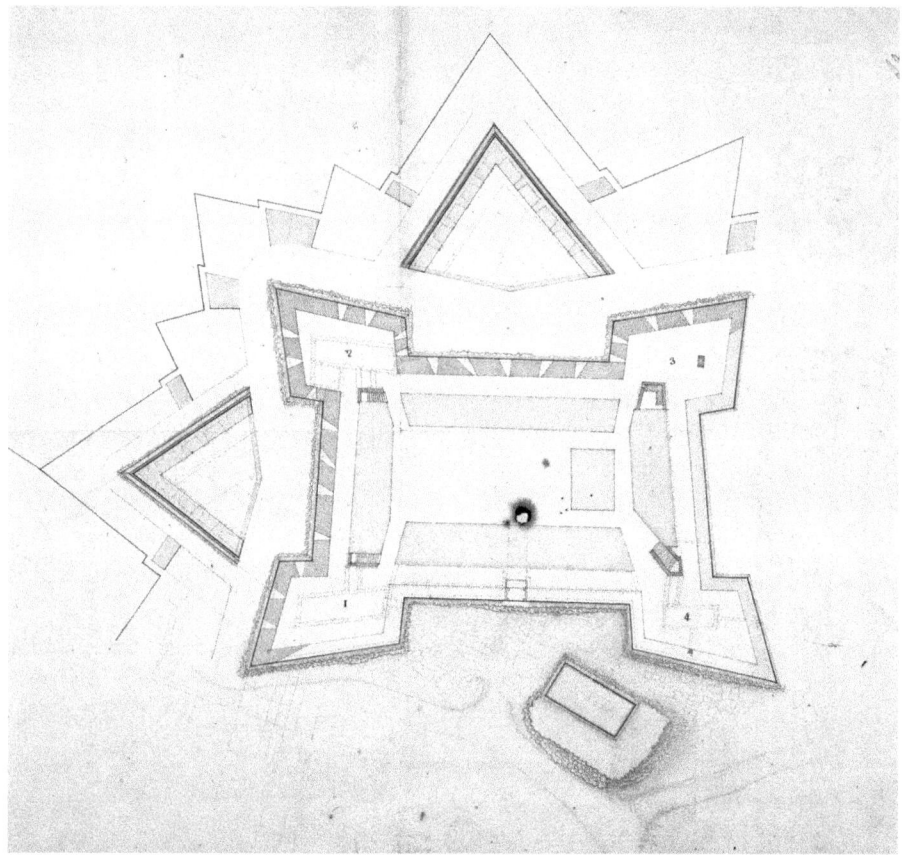

An anonymous French plan of Fort Carillon, circa 1759. The main gate is on the bottom of the page. *Boston Public Library, Norman B. Leventhal Map Center.*

on Lake George was intent on attacking Fort Carillon, but once again, the British attack failed to materialize.

Work continued on Fort Carillon in the spring of 1757, but the fort was never threatened as Montcalm and seven thousand regulars, Canadians and Indian allies used the location as a staging point for the attack on Fort William Henry at the head of Lake George.

The year 1758 would prove a completely different matter for Fort Carillon and its defenders. British plans for the year called for the new commander in chief in North America, Major General James Abercromby, to personally lead an attack on Fort Carillon with 25,000 colonials and British Redcoats. While Abercromby never saw that many troops, by July 1758, he did have 6,300 regulars and 10,000 colonials encamped near the headwaters of

Lake George. When Abercromby launched his rafts, small boats and even a fourteen-gun sloop onto the lake to carry the army forward, the procession was strung out for miles along the mountain waterway.

The British plan, crafted by Abercromby's second in command and the tactical leader of the army, Brigadier General Lord Augustus Howe, called for a landing on the west shore of Lake George and a march through the forest around the rapids and waterfalls that transfer the contents of Lake George into Lake Champlain. While seemingly an ambitious plan, there was merit to the idea. If Howe could appear on the heights above Fort Carillon before the French could fortify this position then the latter would have no choice but to fall back to the fort and likely withdraw farther down the lake once British gun batteries were erected along the ridge. If Howe was extremely lucky, he would appear on the heights while the bulk of the French army was still on the south side of the rapids waiting to contest the road from the portage to the sawmill. If this were the case, the French would be hopelessly trapped between British forces. In both scenarios, Fort Ticonderoga would fall, for once in control of the heights above the fort, the impressive British artillery train would be able to rain havoc down on the structure.

Abercromby's army encountered little in the way of opposition when it landed, and it appeared that it had caught the enemy off guard. Howe soon gathered together a select force of regulars and colonials and pushed into the woods along a crude Indian trail that wound its way around the rapids. It was a difficult march through dense virgin forest made worse when the advanced guard of the British column ran into a lost French scouting detachment. The running skirmish was a disaster for the French, with most of the 350 men in the force either killed or wounded. British losses were light with one exception; Lord Howe, the heart and tactical head of the campaign, was killed in the opening salvo.

With Howe's death, confusion took hold of the British high command. The army continued its march until a major friendly fire incident in the fading light forced Abercromby to call a halt. Some of the troops returned to the landing site, while the rest spent an uneasy night in the forest before returning the next morning. The delays from this event and the indecision that followed were such that by the time Abercromby's army reached the heights above Fort Carillon, it found the French had cut down the trees Lotbiniere had purposely spared and converted them into an eight-foot-tall log wall that now crowned the heights. To make matters worse, Abercromby's scouts reported that the enemy had placed a fifty-yard-deep *cheval-de-frise* of

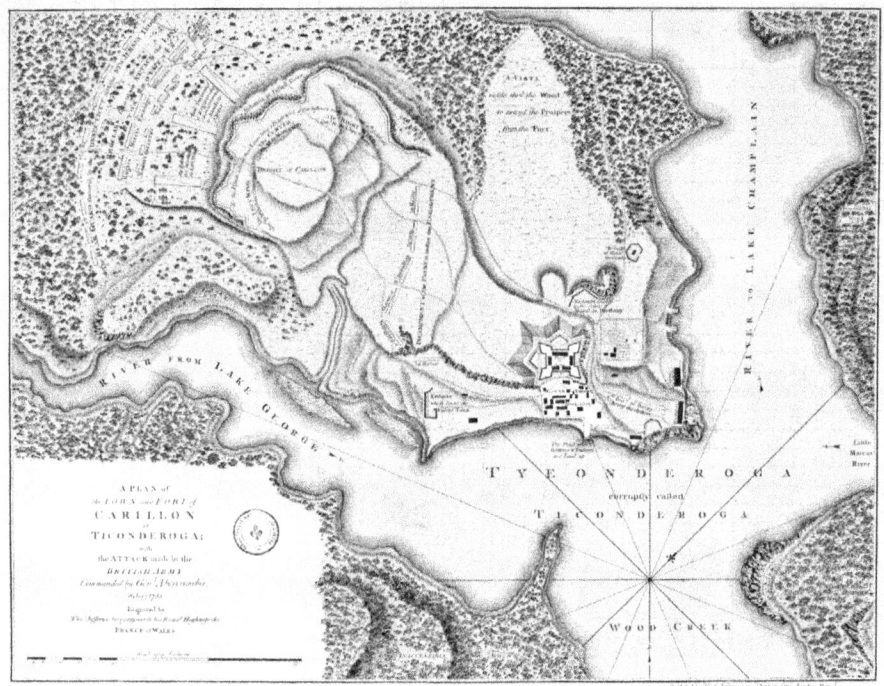

Thomas Jeffery's plan of the Battle of Fort Ticonderoga showing the initial British attack on the French lines above the fort. *Boston Public Library, Norman B. Leventhal Map Center.*

fallen logs and sharpened branches in front of their works, which would make any advance against the position a slow-going affair.

With a superior train of artillery and a plethora of manpower, it appeared that Abercromby would take a more methodical approach toward the fort and spend the next few days bringing his heavy guns and mortars up from the landing site. Once in place, they would quickly demolish the French fieldworks across the heights. The French army would have no choice but to retreat down the lake, and the fort would fall after a conventional siege.

Abercromby, however, was not only distrustful of his colonial troops' fighting abilities but was also convinced that massive French reinforcements would be arriving any day. Fearing that the moment would be lost, and informed by his chief engineer, Lieutenant Matthew Clerk, that the French defenses were flimsy at best, the British general ordered a general assault on the French lines. To support this attack, a battery of guns would be carried by raft to a position at the foot of Mount Defiance to enfilade the makeshift French lines. The latter part of the plan, and the key to the venture's success, failed to materialize as the artillery rafts could not find their landing zone and came under fire from Fort Carillon. The infantry attack had started prematurely, and Abercromby was about to call off the operation when news reached him that the Forty-Second Highlanders had pierced the lines on the French right. Although the general couldn't confirm the reports, he allowed the attack to continue in hopes that they were correct, and in doing so, he gave up control of the engagement.

Waves of Redcoats interspersed with more colorful colonial jackets merged in the wooden maze in front of the French lines. Three times the British regrouped and charged again, but it was a fool's errand. "We were so entangled in the branches of the felled trees that we could not possibly advance," one English officer wrote of the event, while another English officer summed up the effort saying, "It was not in the power of courage or even chance to bestow success." Mercifully, Abercromby sounded the retreat around 6:00 p.m. The British army, having sustained nearly two thousand casualties, spent the night lying on their arms a little over a musket shot from the French lines.[51]

At daybreak, many were shocked when Abercromby ordered the army back to the boats. The command, which several officers attempted to dissuade Abercromby from issuing, set off a wave of confusion. The portage road soon became clogged with men stumbling over one another in the early morning light, dragging supply carts, wounded and artillery on a path that was so muddy that in many places they sank up to their knees. At times, the procession slowed to such a point that exhausted men fell out of the ranks to catch a few hours' sleep, while others, tired of their burden, freely cast aside supplies and nonessentials to ease their plight. Guards had to be posted at the bridge over the La Chute River to control the exodus and prevent the withdrawal from becoming a panicked retreat, which only served to slow the march down even further.[52]

Around mid-morning, the order was given for the army to embark. One colonial soldier referred to it as "a sea of confusion" as soldiers attempted to locate their regiments' boats and, when they couldn't find them, boarded the nearest available vessels. The artillery and wounded were successfully loaded aboard the boats, but in the chaos, much of the supplies were simply abandoned or destroyed to expedite the process. Several British officers attempted to head off the chaos, but with limited success. Around ten o'clock, Colonel David Wooster and Robert Rogers's men suddenly appeared on the scene and were mistaken for the enemy. The confusion now turned to panic as the cries of "Push off! Push off!" echoed down the beach. Fortunately, no

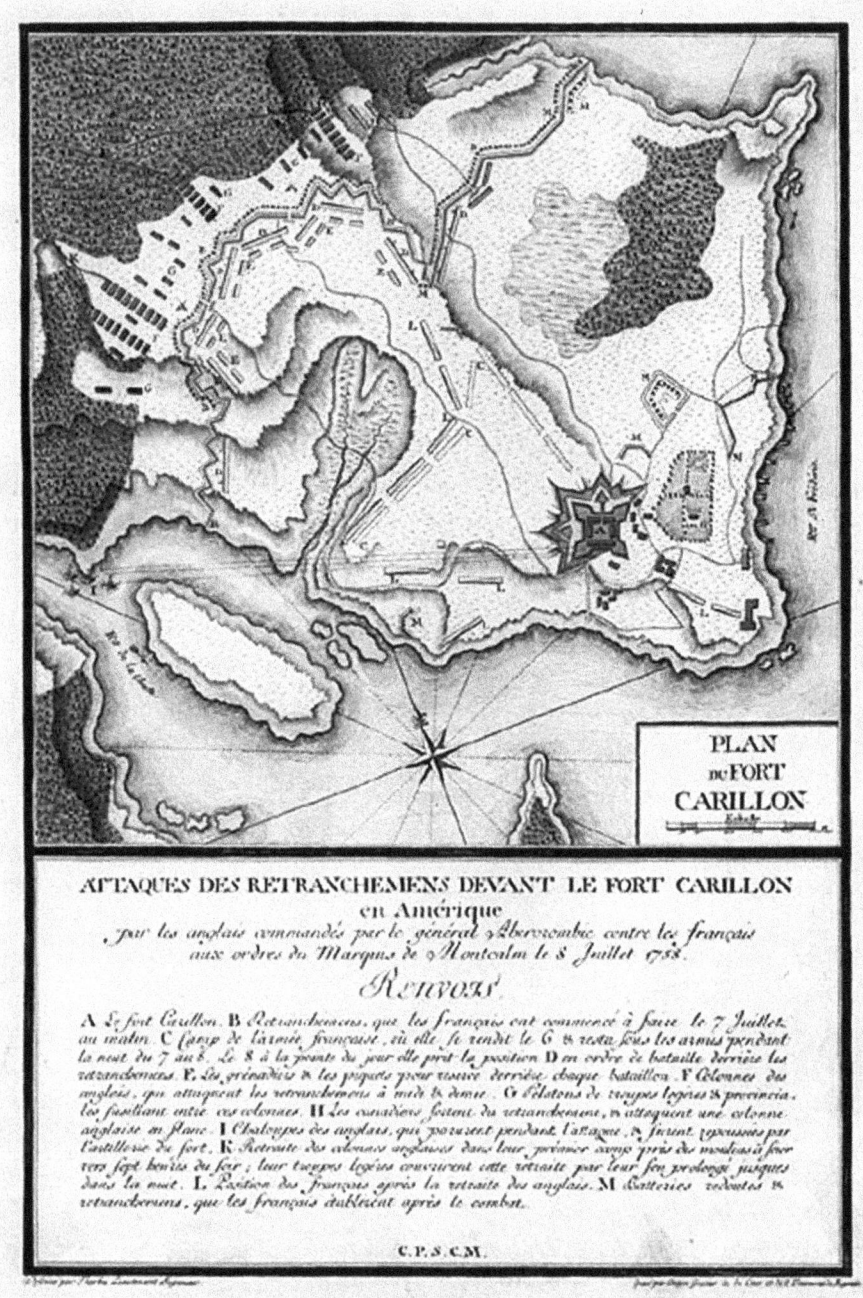

A French map of the Battle of Ticonderoga. *Library of Congress, Geography and Map Division.*

shots were fired, and the error was realized a few minutes later, but by then the inertia of the retreat was too much to overcome. The outlet of Lake George soon became an entwined mass of vessels, each looking to outpace the other down the length of the lake.[53]

As it was to turn out, the fate of both Fort Carillon and Fort St. Frederic was determined long before the British could organize their next expedition. At the start of 1759, the leadership of New France was forced into a decision. With dwindling resources and a lack of manpower, it was simply impossible to guard all the avenues into Canada. As such, it was agreed to abandon both Fort Carillon and Fort St. Frederic once the British, who were certain to return, seriously threatened the positions. Bourlamaque, in command at Fort Carillon with some three thousand men, was to delay and harass the British, but once the outcome was clear, he was to blow up both forts and retreat down the lake to Île aux Noix, where a series of defensive works was being erected.

The new British commander in chief in North America, Major General Jeffery Amherst, was tasked with the 1759 British effort against the French forts in the Champlain Valley and the ultimate aim of seizing Montreal. Amherst, who had overseen the successful capture of the French fortress of Louisbourg the year before, was a much different general than his predecessor and had no intentions of repeating his mistakes.

From the onset, the eleven-thousand-man expedition was focused on a formal siege of the French stronghold at Ticonderoga. Fortune was on Amherst's side. Beyond a gale that threatened Amherst's vessels on Lake George, the army encountered little in the way of opposition when they landed on the morning of July 22 near the French portage on the east side of the lake. By the end of the day, the British force had seized the sawmills and the high ground on the north bank of the rapids. The following morning, colonial detachments probed toward the heights of Carillon, where, to their delight, they found the French had abandoned the entrenchments they had constructed the year before. Thus far, it had been a textbook operation. All that remained now was to start the siege trenches and bring up the heavy guns.

For Bourlamaque, little had gone by the book. With a little over three thousand men at his disposal, the French commander did not have enough troops to man Montcalm's old entrenchments. The only real chance to halt the British would have been to contest their landing, but Bourlamaque had neither enough men to throw the enemy back into their boats nor could he risk them on such a venture. While the enemy dug their trenches and

Major General Jeffery Amherst. A methodical general, Amherst was often criticized as being too slow. The troops serving under him, however, would disagree. Whenever possible choosing maneuver and superior logistics over direct confrontation, the general avoided the costly casualties and failures of his predecessors and, in the process, gained a reputation as a leader who would not recklessly risk his men's lives. *Library of Congress, Prints and Photographs Division.*

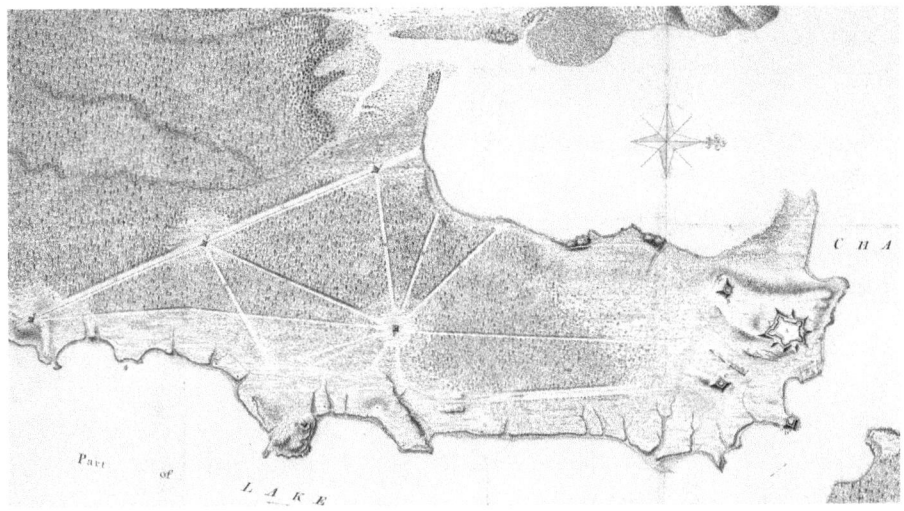

A plan of the fortifications at Crown Point in 1759 by William Brasier. The depth and extent of the defenses envisioned in the plan would make Crown Point one of the most heavily defended places in North America, at least for the last few years of the French and Indian War. The fort and its outworks were never finished as the budget cuts and reductions that came with the conclusion of the hostilities left only skeleton crews to man the fort. *Boston Public Library, Norman B. Leventhal Map Center.*

erected firing platforms on the heights above the fort, Bourlamaque ordered a general withdrawal to Fort St. Frederic. Four hundred men under Captain Louis-Philippe Hebecourt were left behind to continue cannonading the British lines, and the captain was given strict orders to blow up the fort once the British siege guns were in position. True to his orders, Hebecourt kept up a steady fire on Amherst's troops, but by the night of July 26, it was clear that the time had come to abandon the post. Under the cover of darkness and a prevailing fog, Hebecourt's men set explosives in the fort's magazine and pushed their boats out onto the lake.[54]

Around 10:00 p.m., several French deserters appeared before Amherst with news that the French had rigged the fort for demolition and retired down the lake. Amherst called on volunteers to enter the fort and disarm the French charges, but it was already too late. Fires could be seen breaking out inside the fort, and a few minutes later, a tremendous explosion shook the peninsula, sending a shower of debris raining down on the lake and the surrounding countryside. Although the explosion had only destroyed one of the bastions and sections of the connecting curtain walls, the fort's barracks and wooden structures were now fully consumed by flames. Soon, smaller explosions wracked the fort's interior as the garrison's cannons, loaded up to their muzzles with powder and shot and their fuses set, began bursting, each concluding in an eerie cadence of high-pitched ricochets as their deadly contents splattered against the stone walls, making it all but impossible for anyone to approach the fort. Amherst ordered Rogers and the light infantry to pursue the retreating garrison. A few abandoned bateaux loaded with powder and baggage were seized, and a detachment of fifteen Frenchmen was captured when they accidentally wandered into the British lines, but otherwise the French had made good their escape.[55]

On the morning of July 31, French royal engineer Captain Jean-Nicolas Desandrouins informed Bourlamaque that Fort St. Frederic was ready for demolition. With his army riding quietly aboard their small vessels in Bulwagga Bay, the general gave the order to light the fuses. At first nothing happened, but a volunteer was sent forward, who reprimed the mines. Moments later, an explosion rocked the citadel, collapsing it in a sheet of flames. For several long moments, a trance held the soldiers of New France as each sought to put meaning to the column of black smoke that clutched the ruins of Fort St. Frederic. Beyond the popping and hissing coming from the wreckage and the gentle lapping of water against the ships' hulls, it was oddly quiet, until finally Bourlamaque gave the signal for the fleet to get underway.[56]

Although Amherst had captured the twin French strongholds on Lake Champlain within the span of a week, there was little he could do to take advantage of the success. The French were known to have a small fleet of warships operating on the lake, meaning that he would have to build a fleet of his own to challenge them before he could risk moving the army forward in its small boats. In the meantime, the general turned his attention to securing both Ticonderoga and Crown Point. To protect the former location, the general ordered the severely damaged Fort Carillon rebuilt and renamed Fort Ticonderoga, after the native name for the peninsula on which it stood. In something of a vindication of Lotbineire's work, Amherst saw nothing wrong with the structure other than it was a little small. "I will repair the Fort upon the same Plan as the Enemy had built it," the British commander wrote in his journal while the last of the fires in the fort were being extinguished. "Which will save great expense & and give no room for the Engineers to exercise their genius which will be much better employed at Crown Point."

It was at this last location that Amherst reserved most of the fortification work. The general was quick to see the advantages in the location. "This is a great post gained," he noted in his journal after his first visit. "[It] secures

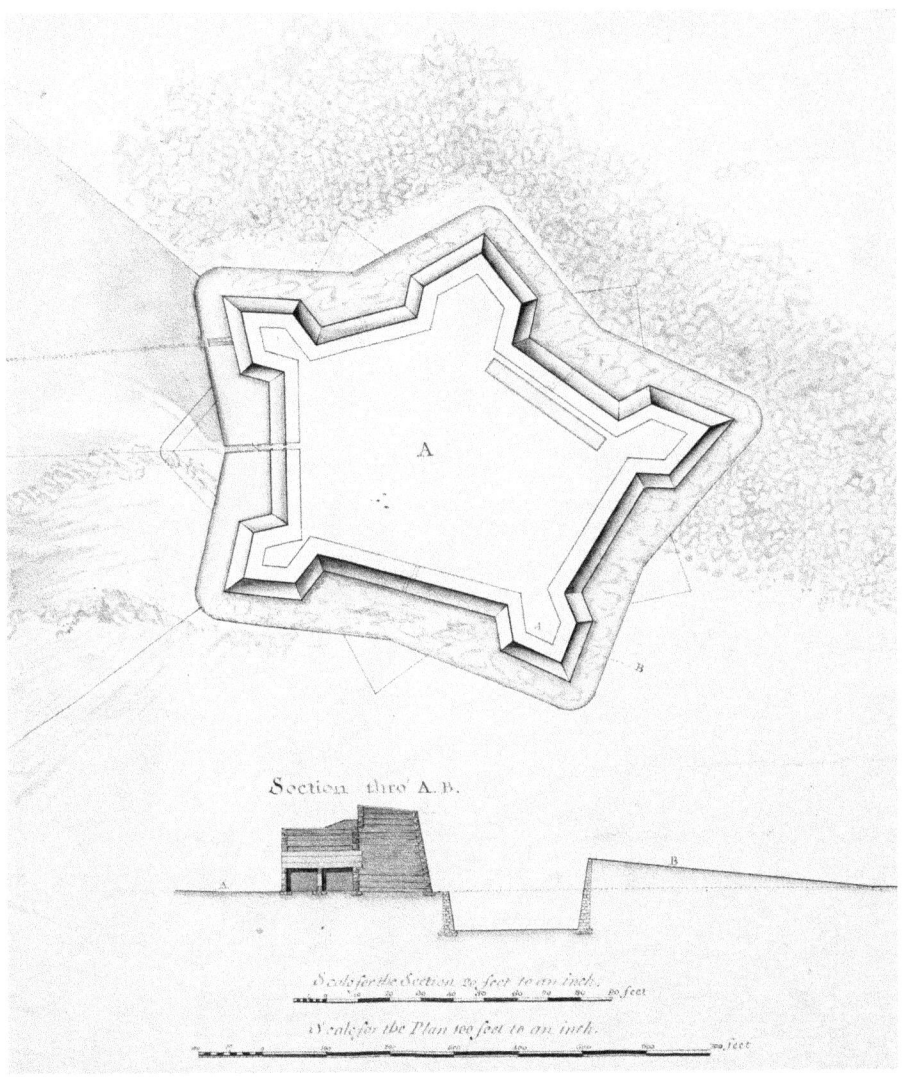

A 1759 plan of Fort Amherst at Crown Point. The plan is likely the work of Lieutenant Colonel William Eyre, who acted as chief engineer for Amherst's campaign in 1759. *Boston Public Library, Norman B. Leventhal Map Center.*

entirely all the country behind it, and the situation and country about is better than anything I have seen." After making the necessary defensive deployments the next day, he surveyed the ground with Lieutenant Colonel William Eyre and selected a location for a fort, which was to be started immediately.

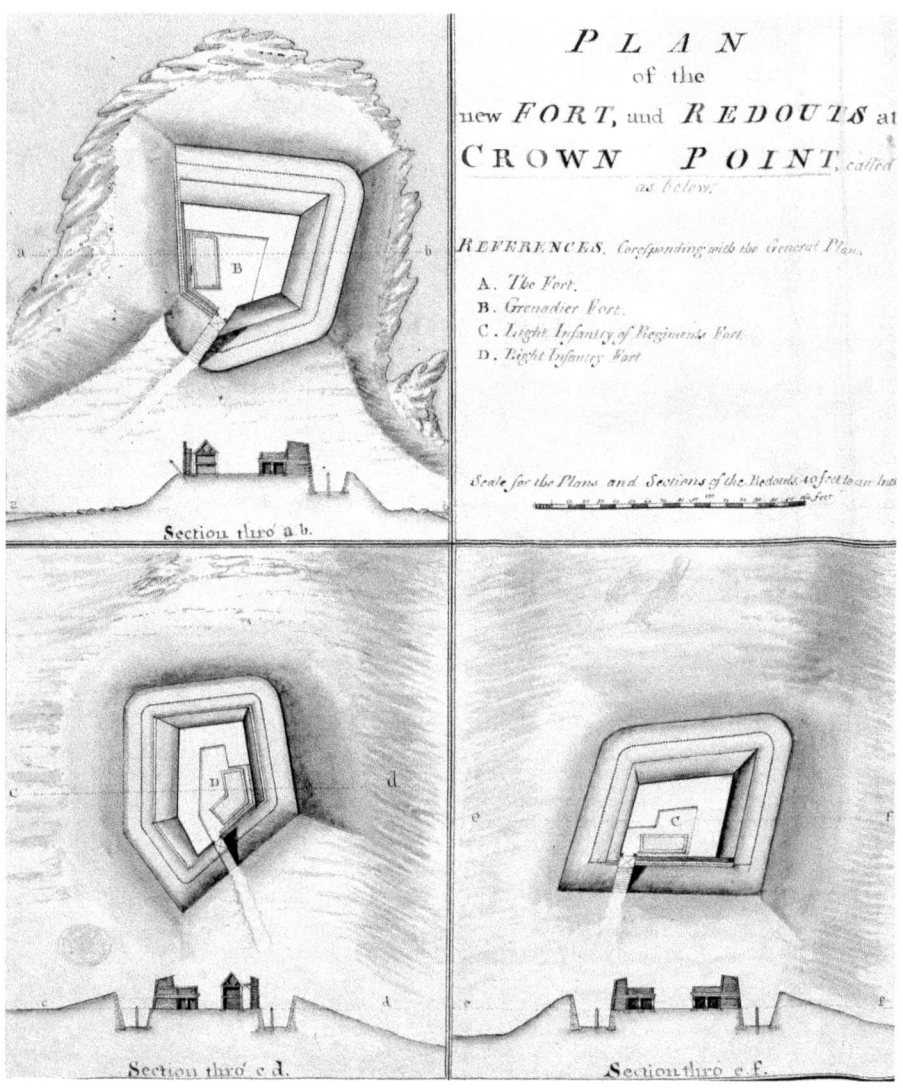

The three redoubts built to support Fort Amherst. *Boston Public Library, Norman B. Leventhal Map Center.*

One thousand men toiled away at the newly laid out Fort Amherst. Set back slightly from the water's edge, it was a sprawling, five-bastioned, pentagon-shaped fort. Capable of holding several thousand men, it was one of the largest forts built in North America for its time. A ditch was dug about the structure and the earth from the effort used to construct twenty-foot-thick walls. These were to have been covered in stone, but

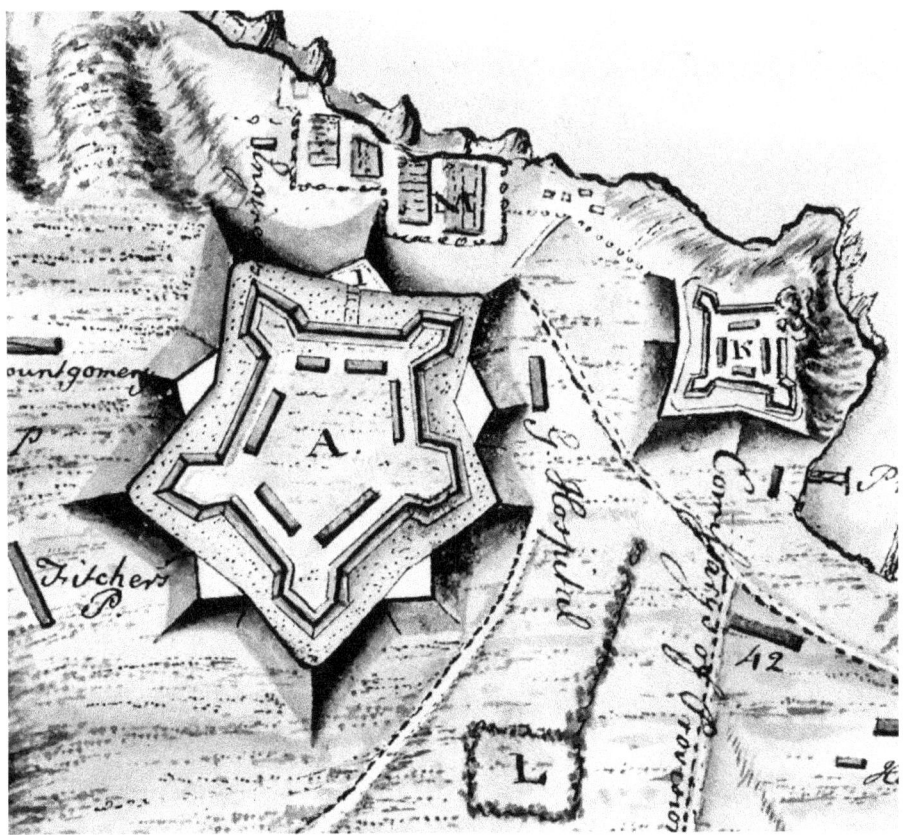

A portion of a larger 1759 map showing Fort Amherst (A) and the ruins of Fort St. Frederic (K). The General Hospital and the encampment for several of the regiments in General Amherst's army can also be discerned. *Library of Congress, Geography and Map Division.*

with the reductions resulting from the end of the French and Indian War, the fort was never completed. Fort Amherst was supported by an extensive network of smaller fortifications. Three stone redoubts known as the Regiment's, Gage's (Light Infantry) and the Grenadier's flanked the fort at the distance of a few hundred yards. Armed with cannons, each was a small fort unto itself. Farther south, the peninsula was cut by another defensive line of blockhouses, and between this and the fort were well-cut roads to allow for the rapid movement of troops. When these elements were put together, it made Crown Point one of the most heavily defended positions in North America.

While Amherst's campaign managed to wrestle naval control of the lake away from the French, it ultimately went no farther than Crown Point, as the rainy fall weather ended the campaign season. Both Fort Amherst and Fort Ticonderoga would be used as staging points for the British campaign against Île aux Noix in 1760, and both would be occupied by a significant number of troops up to the surrender of Canada on September 8, 1760.

With the formal conclusion of hostilities in 1763, the two forts were reduced to small garrisons, mainly occupied in customs and maintenance duties. Fort Ticonderoga would go on to play an important role in the American Revolution, but Fort Amherst's fate was sealed by burning embers from a fireplace on a windy evening in 1773. One of these embers found the fort's powder magazine, and the resulting explosion, which was clearly heard at Ticonderoga, was powerful enough to rain fire and debris on and around the fort. The damage from the blast and the resulting fires was severe enough that the fort was never repaired. Although a handful of men were stationed at the post to secure the cannons stored there, for all practical purposes Fort Amherst ceased to be a fortification.

Chapter 3

LAKE GEORGE AND THE UPPER HUDSON VALLEY

The forts of the upper Hudson Valley all trace their roots to the original forts at Albany. First among these was a Dutch trading post erected on Castle Island in 1614. Named Fort Nassau, the post was constructed like many of the early forts of North America, by men with little or no formal training in the art of military fortification. The site was cleared of debris, and then a series of deep ditches were dug that traced out the position of the fort's walls. Freshly cut logs were placed vertically in the ditches and lashed together, and the trenches were then filled in with burnt earth and gravel to support the entire structure.[57]

In its completed form, the stronghold was nothing more than a square palisade, measuring fifty feet to a side with an eighteen-foot-wide moat dug about its perimeter and a drawbridge at the main gate. Inside this stockade sat a thirty-six-by-twenty-six-foot trading house that not only harbored the trade goods but also acted as the living quarters for the dozen or so traders who occupied the post. A pair of light cannons sat within the fort's parade ground for defense, while eleven *pierriers* (small swivel guns designed to fire rock projectiles) lined the structure's walls. As it was, Fort Nassau's days were numbered. The low-lying island, while seemingly an excellent defensive choice, was subject to the annual ravages of the Hudson River. In the spring of 1617, the fort was so badly damaged by floodwaters that its occupants were forced to abandon the post and build a new one on the west bank of the river a few miles south of the old site.[58]

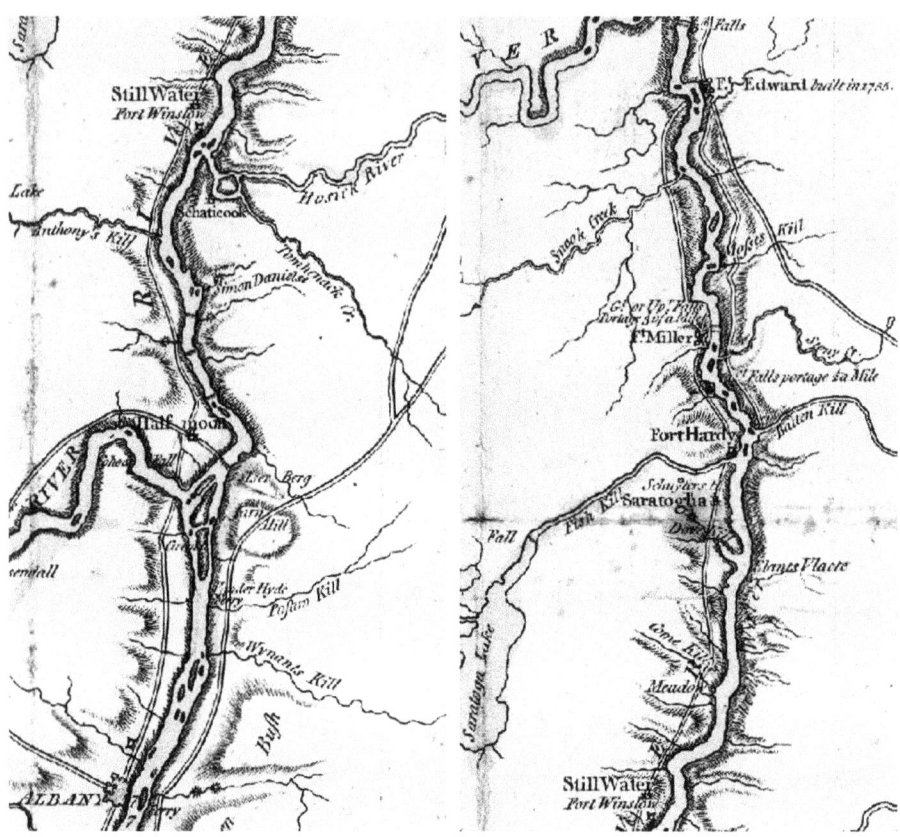

Left: A colonial map of the Hudson River from Albany to Stillwater, and *right*: from Stillwater to Fort Edward. Note the falls/rapids marked on the map near Half-Moon and Fort Miller. *Library of Congress, Geography and Map Division.*

In 1624, the second Fort Nassau was replaced by a larger structure to better accommodate the burgeoning fur trade. Fort Orange, as it was called, was moved back up the river almost opposite the old Fort Nassau on Castle Island. A four-bastioned structure, measuring 150 feet to a side, the fort was of earth and wood construction. A pair of 15-foot parallel walls, made in the same fashion as Fort Nassau, were erected and traced out the fort's perimeter. The walls were then braced with cross members at regular intervals and the intervening space between them filled in with earth, most likely taken from the moat that surrounded the fort on three sides. A rampart or walkway was then fashioned in the area between the two walls by placing planks on horizontal cross members, which were then secured to the palisades on either side. It is not clear how thick the walls of Fort Orange were, but given that one citizen who had a dwelling within the fort petitioned to cut a door in the curtain wall to allow easier access to his home, one suspects that they were thinner rather than thicker.[59] Protruding diamond-shaped bastions were placed at each corner of the structure to allow the defenders to sweep the walls with gunfire and served as firing platforms for the fort's eight large stone firing guns.

The new fort was clearly an improvement over Fort Nassau, but earth forts had inherent drawbacks, and these plagued Fort Orange from the outset. Foremost, earth forts constructed in this fashion tended to lose the earth packed between the walls through the space between the vertical logs that made up the palisade. In addition to this basic problem, the palisades themselves tended to rot away after a few years. Seasonal extremes and substantial rainfall, both of which were present at Albany, accelerated this process, and if the builders were not careful in sealing the structure, leaks could result in standing pools of water that magnified the problem. This last issue was typically more pronounced with the casemates under the bastions where the garrison would store their powder, munitions and foodstuffs, all of which, in turn, suffered. These shortcomings were summarized over a century later by British general John Campbell, the Earl of Loudoun. Loudoun would write a colleague about Fort William Henry, also an earthen fort:

> *In the works that were carried on there last year, that the timber has already suffered; and in the Casemates there, where the Water has Soaked through; but the great Logs, from not being sufficiently secured with Oakum, are very much Rotted; and even the People here, agree that the Timber of this Country, rots much sooner than the Timber in Europe does; but indeed there is no Justice done to it here, for it is cut when wanted, and directly put to use, whatever the Season of the Year is; For which reason, whenever there is occasion to build a Fort, that probably will remain, if there is Stone & Lime near, I should advise it's being built of them.*[60]

It seems likely that the forty or so traders and soldiers who garrisoned Fort Orange quickly discovered the fort's shortcomings over the intervening years, but with limited resources, many of the problems were simply overlooked, leading one observer in the late 1640s to refer to the location as the "miserable little fort called Fort Orange, built of logs."[61]

In 1664, Fort Orange faced its only real test. In late September, an English frigate carrying Colonel George Cartwright dropped anchor before the fort and demanded its surrender. Cartwright's appearance was not wholly unexpected, as news had already reached the garrison that a English task force had captured New Amsterdam. Outgunned, with a decaying stronghold around him and no prospects of relief, the commander of Fort Orange wisely chose to accept the Englishman's terms and surrendered the fort without firing a shot.[62]

With the change of ownership came a reassessment of the town of Orange and its fort. The first was renamed Albany in honor of James II, the Duke of Albany, while the second, after a decade of occupation, was abandoned in favor of another stronghold better positioned to protect the growing town. Constructed in the spring of 1676, Fort Albany, as it was called, was placed on a hill to the northwest of the city. From this new vantage point, the fort's cannons could cover the town, its main approaches and the river. High ground to the west called the position of the fort into question, but as it was unlikely that cannons could be dragged through the wilderness to take advantage of this weakness, this had little influence on the fort's construction.

Fort Albany's main walls were constructed of fifteen-foot-tall pine trees anchored and lashed together in a typical stockade fashion. A second shorter and more widely spaced row of vertical logs backed the main walls and acted as supports for the platforms and walkways that circled the inner perimeter. Bastions were added at each corner and planked over to allow for the mounting of five or six small cannons. The northwestern bastion was later rebuilt in stone and acted as the fort's magazine. Within the compound, two long three-story buildings were constructed along the north and south curtain walls, both of which were tall enough that their second-story windows overlooked the main walls. A ditch backed by sharpened stakes circled the stronghold on three sides to provide protection against a surprise attack.[63]

Garrisoned by one hundred men and armed with twelve light cannons, Fort Albany was sufficient protection against French and Indian marauders, but it failed to address the needs of the town to protect itself from these types of attacks. To deal with this problem, work turned toward enclosing the town within a stockade that was supported at key points by log blockhouses. Benjamin Wadsworth, a Massachusetts representative who traveled to Albany in the early 1690s, left a description of this work:

> *Ye town is incompass'd with a fortification, consisting of pine-logs, ye most of ye a foot thro, or more; ye are hewed on two sides and set close together, standing about 8 or 10 foot above ground, sharpened at ye tops. There are 6 gates; 2 of ye east to ye river, 3 north, one south; ye are five block-houses; 2 north, by two of ye foremention'd gates, and 3 south.*[64]

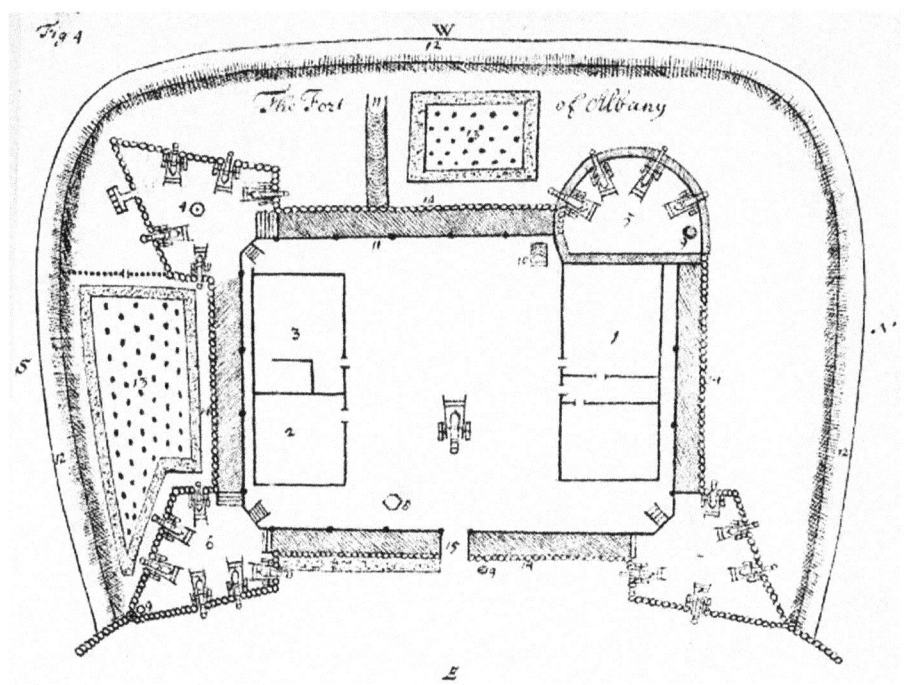

A plan of Fort Albany, 1695. *From Miller,* Description of the City and Province of New York, *1695.*

Legend:

1. The governor of Albany's house.
2. The officers' lodgings.
3. Soldiers' lodgings.
4. The flagstaff and mount.
5. The magazine.
6. The dial mount.
7. The town mount.
8. The well.
9. The sentry boxes.
10. The magazine.
11. The sally port.
12. The ditch with stakes.
13. The gardens.
14. The Stockado.
15. The fort gate.

Although simple and expedient, the wooden Fort Albany, like Fort Orange, suffered the wrath of the elements in the years following its construction. The horizontal planks that made up the flooring for the bastions were particularly vulnerable, having to be replaced just a few years after their installment. The rest of the fort soon followed, and by 1687, Governor Thomas Dongan was writing England that

> at Albany there is a Fort made of Pine Trees fifteen foot high & foot over with Batterys and conveniences made for men to walk about, where are nine

guns, small arms for forty men four Barils of Powder with great and small shott in proportion. The Timber and Boards being rotten were renewed this year. In my opinion it were better that Fort were built up of Stone & Lime which will not be double the charge of this years repair which yet will not last above 6 or 7 years before it will require the like again whereas on the contrary were it built of Lime and Stone it may be far more easily maintained.[65]

Fort Albany and the surrounding town were alarmed several times during King William's War (1689–97), but a serious attack on the town never materialized. Even so, several towns in the area were ravaged during the conflict, and it was clear that if something was not done about the defense of Albany and the surrounding communities, a major French effort in the future could collapse the entire New York frontier, taking the English and their Iroquois allies with it.

To help deal with the problem, a senior royal engineer, Colonel Wolfgang Romer, accompanied the new governor of New York and New England, Richard Coote, Lord Bellomont, to the colonies in 1698. Although peace reigned in Europe at the moment, signs were clear that it would not last, and as such, one of Bellomont's primary tasks was to assess and repair the colonial defense of New York and New England. In pursuit of this goal, shortly after their arrival, Bellomont directed Romer to conduct a survey of the fortifications of New York, starting with those of the upper Hudson Valley. The governor was anxious to know what it would take to secure this vital frontier and with it the allegiance of the Iroquois Nations, without whose help New York would find itself in dire straits should a future conflict erupt.

Romer spent the next several months examining Albany and the nearby posts and towns. He quickly grasped the importance of the area as a steppingstone for the French into New England, New York and New Jersey should they make a concerted effort to seize it. And they might well try, given that in the engineer's opinion, the current defenses were laid out without guidance or thought to the actual security of the frontier. "It is a pity, and even a shame, to behold a frontier neglected as we now perceive this is," he informed Bellomont in one of his early reports. "Had the public interest been heretofore preferred to individual and private profit, which has been scattered among a handful of people with diabolical profusion, the enemy would had never committed pernicious forays on the honest inhabitants."[66]

Lake George and the Upper Hudson Valley

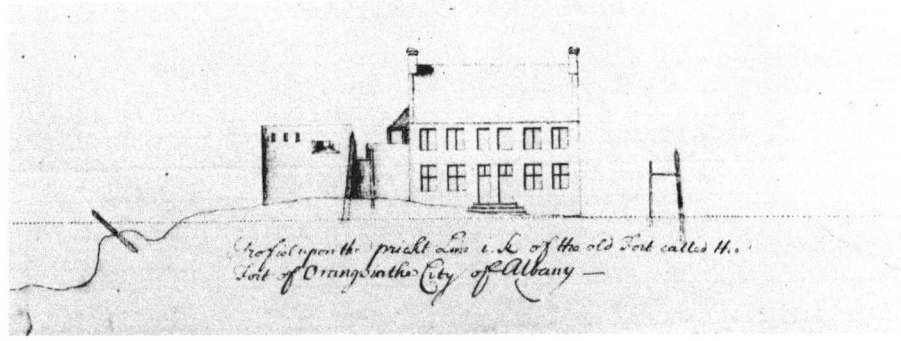

A profile of Fort Albany, circa 1698, by Wolfgang Romer showing the northwest bastion, the governor's house, a cross-section of the fort's curtain wall and the cheval-de-frise that circled the fort. It is interesting to note that Romer refers to the fortification as the Fort of Orange, which, although incorrect, may speak to how the town's inhabitants referred to the fort at the time. *From Hulbert,* The Crown Collection of Photographs of American Maps, *Series I (1907).*

In his report, the engineer identified five locations that required attention. The two most important were Schenectady and Albany. He recommended that poorly laid out wooden palisade structures at these locations be replaced with regular stone forts. In addition, he also pointed to the need to fortify two northern locations: Half Moon, located at the junction of the Mohawk and Hudson Rivers a dozen miles north of Albany, and Conestoga, about the same distance to the northwest along the south bank of the Mohawk River. These points, protected by a "good guard house or stone redoubt" and garrisoned with thirty or forty men, would act as barriers against French excursions and points of refuge for settlers of the region. Lastly, Romer gave some consideration to the hamlet of Saratoga, located almost thirty miles north of Half Moon on the east bank of the Hudson River. The small settlement here had been all but abandoned during the last conflict. As the post secured English claims to the region, he suggested that a palisade fort with a small stone tower at its center be constructed to encourage settlers to return and offer them some form of protection should French raiders appear before the place.[67]

While Bellomont was pleased with the report, little was done to act on Romer's suggestions. Work was started on a stone fort at Albany, but it was never finished due to lack of funds. The proposed stone redoubts at Half Moon and Nustigione never materialized. A pair of old palisade forts at these locations were revived at various points during the intervening years, but in general, both sites were neglected to the point of decay.

The outbreak of Queen Anne's War (1702–13) would once again focus English martial efforts in this region. In 1709, a plan was put forward to launch a pair of expeditions against Canada. One component of the joint attack, a fleet carrying five regiments of regulars, was to arrive from England, and after taking on supplies and additional colonial troops in Boston, it was to sail down the St. Lawrence River and besiege Quebec. The second element of the plan called for a small force under General Francis Nicholson to assemble at Wood Creek and, from there, proceed down Lake Champlain to besiege Fort Chambly and threaten Montreal. The consensus among the planners was that the joint effort would divide the forces of New France, making the seizure of the French colony a real possibility.

To accomplish his portion of the plan, Nicholson was forced to build a supply line from Albany to an encampment on Wood Creek, close to fifty miles away. It was a daunting task that spawned a series of palisade forts and wooden storehouses extending from Fort Albany to the newly built Fort Schuyler at the head of Wood Creek. In its final form, supplies were loaded into canoes at Albany and sailed ten miles upriver to Van Skaiks Island near the lower confluence of the Mohawk River. Here the water became so shallow that the canoes had to be brought ashore and carried overland to a storehouse built at Half Moon, a peninsula formed by the junction of the Mohawk and Hudson Rivers. The supplies were then loaded onto wagons and carried twelve miles north to Stillwater, where the army had built a palisade stronghold called Fort Ingoldsby.

There, the supplies were placed back into boats and taken eighteen miles upriver to a point called the First Carrying Place. A set of shallow rapids at this location often forced the crews to drag their canoes through armpit-high water to reach the staging point on shore. The canoes were then unloaded and their contents carried seven hundred yards around the intervening rapids. They were then reloaded into another set of canoes and transported four miles to the Second Carrying Place. Once again, the crews emptied their vessels and dragged their contents eight hundred yards around the shallow rapids. There, the supplies were loaded into a final set of canoes and sent sixteen miles upriver to the Great Carrying Place, where the army had built another palisade fort called Fort Nicholson. From there, wagon teams carried the supplies to a set of storehouses halfway up the recently cleared trail to Wood Creek. Teams of wagons sent from Fort Schuyler were dispatched to these storehouses to cover the remaining eight miles.

Great efforts went into maintaining Nicholson's supply lines, but the campaign of 1709 was abandoned when the British fleet failed to arrive.

LAKE GEORGE AND THE UPPER HUDSON VALLEY

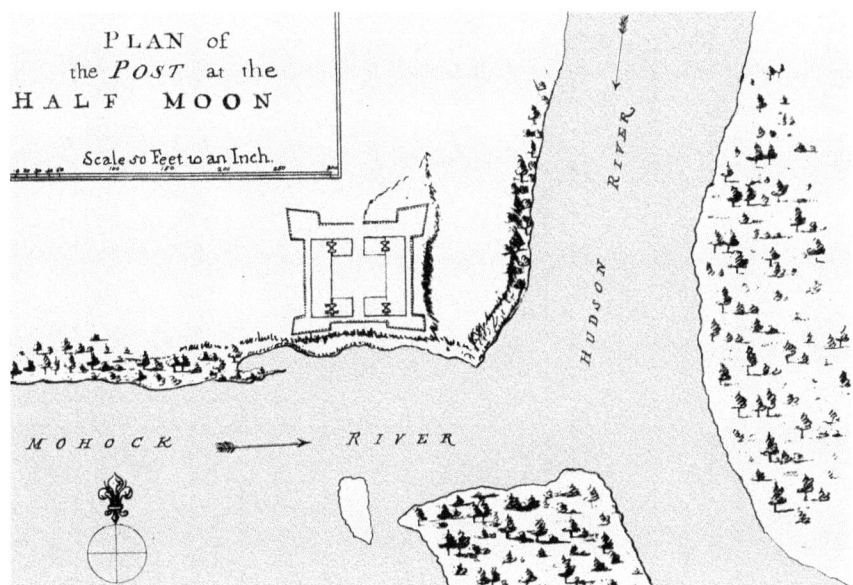

The fortified post at Half Moon, circa 1758. Half Moon was one of the first portages for small boats headed north from Albany. *From Hulbert,* The Crown Collection of Photographs of American Maps, *Series II/2 (1907).*

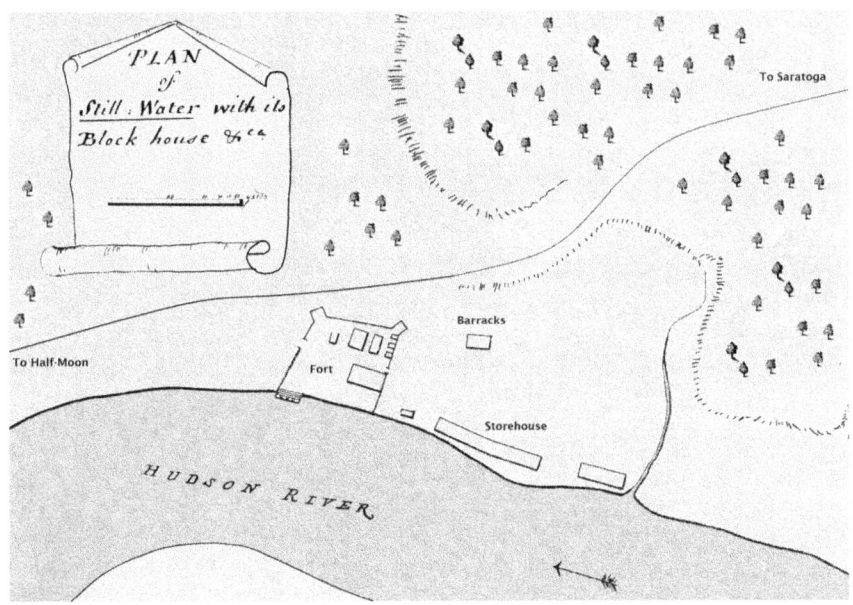

The fortifications at Stillwater. *From Hulbert,* The Crown Collection of Photographs of American Maps, *Series I/3 (1907).*

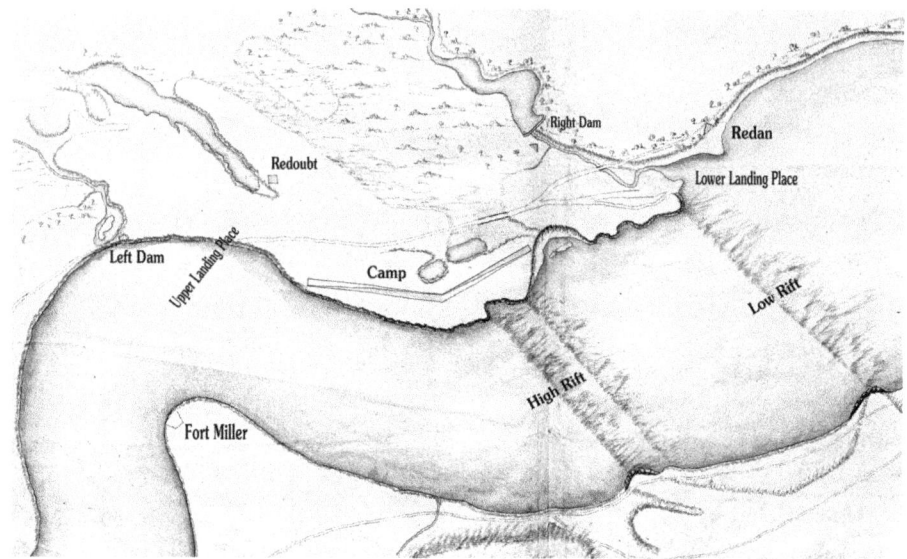

A 1759 map of the Great Falls on the Hudson River. While referred to as the Great Falls in this plan, the area was more commonly known as the Second Falls or Second Carrying Place. The location of Fort Miller on the west bank of the river and several outworks near the landing zone on the east bank can be readily seen. Note that the captions have been modified for clarity and that the river flows from left to right in this image. *Boston Public Library, Norman B. Leventhal Map Center.*

With no money for troops to man them and no desire to let the enemy occupy them, the advanced posts on the upper Hudson were put to the torch. Two years later, the same plan was reenacted, and many of the men who built the original strongholds along the upper Hudson found themselves repairing their old works. This time, a new fort was erected at Wood Creek. Named in honor of the queen, Fort Anne was a simple palisade fort with bastions at the four corners and a number of small buildings clustered within its interior. Just as a few years before, the expedition would prove for naught. The campaign was abandoned when the British fleet wrecked in the St. Lawrence River and turned back for Boston. The response was similar, and once again, columns of smoke dotted the upper Hudson as the colonial troops withdrew.

For nearly a decade after Nicholson's last campaign, there was little interest in the fortifications of the upper Hudson Valley. Such an attitude meshed nicely with the lack of funds to do anything meaningful. In 1720, Governor William Burnet finally took up the matter, informing the Board of Trade that he had levied a 2 percent duty on European goods to raise money toward building new stone forts to replace the wooden ones at Albany and Schenectady. In the meantime, the old wooden fort at Albany had been repaired and new wooden palisades had been erected around it and the town.[68]

At Saratoga, the most northerly defensive point along the Hudson, a small stockade fort was erected in 1721 to provide refuge for the handful of Dutch farmers in the area, but the scarcity of funds and general disinterest in the project resulted in such a poor fortification that the twelve-man garrison often complained that it was impossible to keep arms and supplies dry during rainy weather.

More concerted efforts were made at Schenectady. Portions of the town were rebuilt after a devastating French attack in 1690, but it would be another five years before the wooden palisades that originally encircled the village were reconstructed, this time with blockhouses at various points along their length. Although, much to the citizens' and the governor's dismay, the proposed stone fort was never constructed, in 1705 a one-hundred-square-foot palisade fort with bastions at each corner was built at the eastern edge of the town. Except for frequent repairs, the "Queen's Fort," as it was known, would remain unchanged for the next thirty years.

In 1734, the weather-beaten Queen's Fort was torn down and a new one of the same dimensions erected in its place. Christened Fort Crosby, this new stronghold was far sturdier, being built by laying horizontal timbers one upon the other on a stone foundation as opposed to the vertical post method of the old fort. Capable of holding several hundred men, the new stronghold housed small four-pound cannons in each of the bastions and six- and nine-pound guns on carriages in the parade ground. As the fort was garrisoned by only seventeen men, maintenance issues and a perpetual lack of supplies undermined the overall strength of the works.[69]

In the mid-1730s, efforts were also taken to repair the weather-torn wall and blockhouses that surrounded Albany. A few years later, in an April 1734 address to the assembly of New York, Governor William Crosby called for stone fortifications at Schenectady and Albany. Taking the governor's concerns to heart, the New York assembly responded by passing a bill titled "An Act for Fortifying the City of Albany and Schenectady and Other Places in the County of Albany," which Crosby quickly signed. A year later, work began in Albany on a new stone fort using a thirty-year-old trace laid out during Queen Anne's War. Fort Frederick, as the new structure was eventually named, measured two hundred feet to a side and boasted fourteen-foot-tall walls between its four bastions. A twenty-foot ditch surrounded the structure, and a pair of two-story buildings slightly higher than the fort's walls ran down the length of two sides. These twin structures served as barracks, storehouses, workshops and the commandant's quarters. Armed with two dozen cannons and garrisoned by nearly one hundred men, Fort Frederick was a powerful addition to the defenses of Albany. Reverend Samuel Chandler recorded his impressions of the fort when his regiment passed through Albany in October 1755:

> *The Town or City is picadoed, abt two miles round on the west side, on a High Eminence is a Fort or Citadell: Stone and Lime. Four Bastions acute Angles abt 45°. Two handsome buildings or Barracks. Brick fences but stone on back side. Abt 14 guns, 2 before the Gate; Garrisoned by an independent company of 100 men. Captain Rutherford 15 men mounted upon guard the east side of the fort next the town, abt 24 loop holes upon the parapets.*[70]

In addition to the new fort, a new fifteen-foot-tall palisade was erected around the town, and several blockhouses armed with small cannons were placed at key points along the perimeter. By 1738, all the work was completed, making Albany the most secure post on the New York frontier.

Due to its harbor, Albany would remain a key location during the last two French and Indian Wars (King George's War, 1744–48, and the French and Indian War, 1754–63). The primary focus of English fortifications, however, would shift north in response to a series of major campaigns aimed at seizing the French-held Fort St. Frederic and later Fort Carillon. The old supply route up the Hudson was revived, which not only required replacing and repairing the old works but constructing several new ones as well.

General William Johnson, who would lead a colonial campaign against Fort St. Frederic in 1755, was responsible for a string of these new fortifications. The old fort at Stillwater was replaced and named Fort Winslow after Johnson's second in command. Farther north, Fort Hardy was raised at Saratoga on a hill on the northeast side of the town, and at the Second Falls, a wooden four-bastioned post known as Fort Miller was erected to help cover the portage.[71]

While these posts and a dozen smaller redans and blockhouses provided some measure of security from French and Indian War parties, it was at the north end of this chain of forts that Johnson built a pair of famous strongholds. The first of these was at the Great Carrying Place, where even small boats and canoes could go no farther. Here on the east bank

A portion of a 1756 map of Albany showing the palisade walls and supporting blockhouses that enclosed the town. Fort Frederick can be seen positioned along the northwest edge of the city, while the ruins of Fort Orange are clearly marked out to the south. *From Hulbert,* The Crown Collection of Photographs of American Maps, *Series I/2 (1907).*

of the Hudson, Johnson stopped to construct Fort Edward. The colonial commander was in luck when it came to this endeavor. Early in his campaign, Johnson had asked for a military engineer to help him build defensive posts along his route and oversee the siege of Fort St. Frederic when the army arrived before this post. In response to the request, Captain William Eyre of the royal engineers was assigned to Johnson's staff.

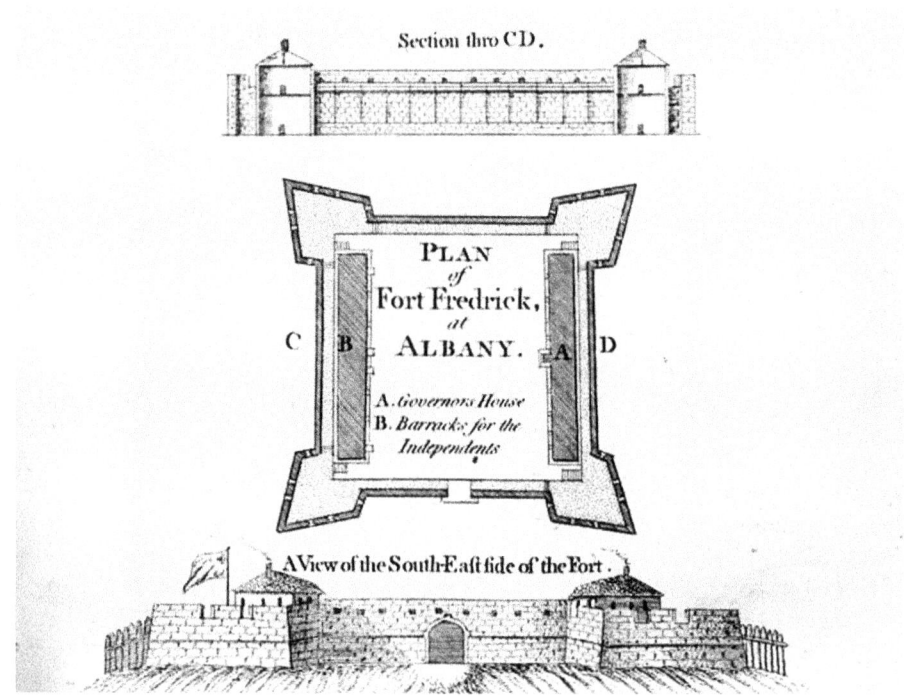

Fort Frederick at Albany, circa 1755. *From Rocque,* A Set of Plans and Forts in North America, *1765.*

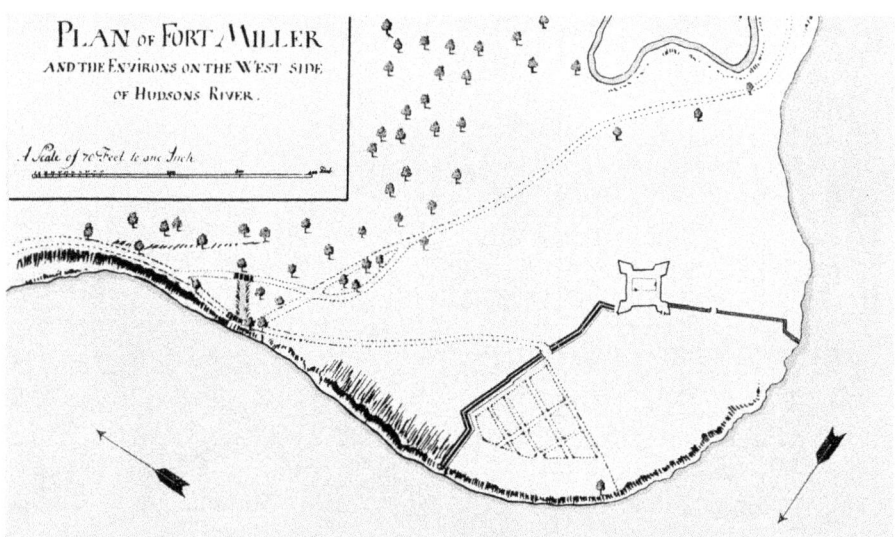

A 1759 plan of Fort Miller at the Second Falls. The fort was erected to guard the important transfer point for supplies moving up the Hudson River. *From Hulbert,* The Crown Collection of Photographs of American Maps, *Series I/3 (1907).*

In positioning Fort Edward, Eyre selected a site along the riverbank that encompassed an abandoned stone structure known as Lydius' Trading Post. Used as the fort's powder magazine, this architectural inclusion made the resulting fort irregular in form. So, too, did the inclusion of a water gate along the fort's southwest corner. A bastion was omitted in this location to accommodate this function, which made moving supplies, men and especially cannons into the fort quicker and safer. Johnson assigned Eyre three hundred men to work on the project. Speed was of the essence. The fort had to be constructed in a few weeks given that the army was preparing to move to the headwaters of Lake George in that time to stage for a final assault against Fort St. Frederic. Eyre, in a letter to Governor Shirley, left an account of Fort Edward not long after its construction:

> *I have built a Fort at the Carrying place, which will contain 300 Men; it's in the form of a Square with three Bastions, & takes in Col Lydius's House; This Work is pallisaded quite round, which is its chief Security from a surprize or sudden Attack; as I was oblig'd to leave that place, and most of the Troops to come here, it was out of my power to make the Rampart and Parapet, of a sufficient height and thickness, to stand Cannon, or the Ditch wide and deep enough to make its Passage very difficult; however I think 3 or 400 Men will be able to resist 1500, provided they do their Duty, if Cannon is not brought against it.*[72]

As it turned out, Eyre did not have time to oversee much of Fort Edward's construction. By the first week of September 1755, the engineer was busy at the base of Lake George laying out another fort while Johnson's army gathered at this location. Nothing more than a small palisade structure, this fortification was to be a safe place of retreat in case operations around Fort St. Frederic went awry.

The engineer had only worked on the new picket fort for a few days when a French and Indian War party under the command of Baron Jean Dieskau attacked Johnson's camp. Although the enemy was defeated, the large number of colonial casualties coupled with the realization that Fort St. Frederic had been recently reinforced by two regiments of French regulars led to sagging morale and a conviction from the army to do no more. With the season progressing and his army unlikely to advance, Johnson turned to building a larger fort at the headwaters of Lake George. The actual construction of what would be Fort William Henry proved more controversial than the general had imagined. His troops showed little interest in building a fort, in

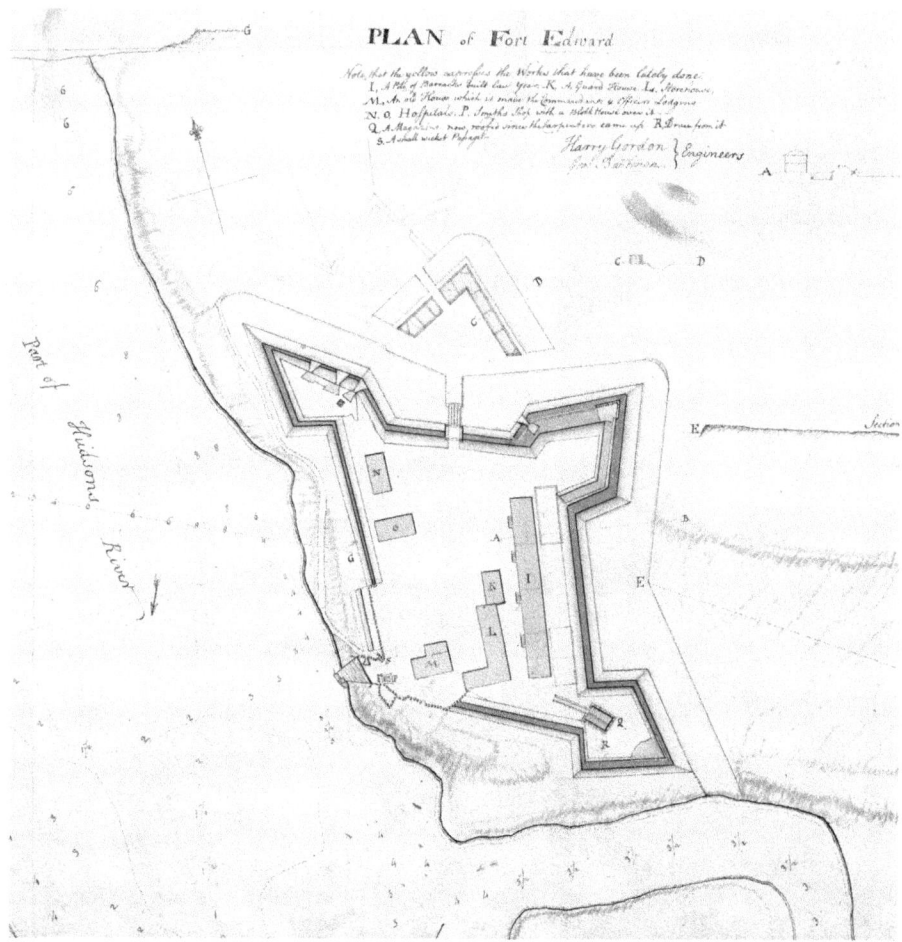

A plan of Fort Edward, circa 1756. *From Hulbert,* The Crown Collection of Photographs of American Maps, *Series II/1 (1910).*

part because it meant the possibility of winter garrison duty. Massachusetts governor William Shirley, the new commander in chief in North America after Braddock's death, expressed his opinion that a fort at the location would be too vulnerable and that it would only cover one of the two possible French and Indian routes into the upper Hudson Valley. Instead, he wanted to see Johnson advance and more efforts made toward securing Fort Edward.

On Johnson's side was Eyre, who, like Johnson, wanted to see a structure able to resist a conventional siege. This meant a fort able to mount and withstand cannons, a task unsuited for the picket fort he had started before the Battle of Lake George. Neither man was foolish enough to propose the

William Johnson. Although Johnson, who would be made a baronet after his victory at Lake George on September 9, 1755, did not accomplish his primary task of seizing Fort St. Frederic, his campaign was responsible for establishing a chain of forts that would ultimately be used by British and colonial forces to drive the French from the Champlain Valley and besiege Montreal. *Johnson Papers, I.*

construction of a structure strong enough to repel a true European-style siege; there were perhaps only two or three of these in all of North America. What they wanted was a place strong enough to bar the advance of any French army bent on invading the upper Hudson and one secure enough to hold out until an army marched to its relief.

Johnson continued to push the matter, and by the end of September, Governor Charles Hardy of New York and Governor Thomas Fitch of Connecticut had thrown their support toward "a respectable fort" at the location. On September 29, 1755, a council of war at Lake George reluctantly agreed that "a place of strength with magazines and storehouses and barracks" would be constructed "with all possible dispatch." The new fort would occupy the site of the current picket one and would be large enough to accommodate a garrison of five hundred men should the need arise. Seven hundred men, excused from all other duties, were detached to work continuously on the project under Captain Eyre's guidance.[73]

Eyre pointed out that the army's current encampment was the best location for a fort. A structure there could protect the boat landing and makeshift docks, as well as cover the recently constructed road to Fort Edward, but the colonial troops were not interested in moving their encampment with another French attack still a possibility and balked at the idea. The engineer's next choice was hardly ideal but was probably the second-best location. He positioned the fort to the northwest of the present encampment on a twenty-foot sand bank along the lake's edge. The lake protected the northeast wall and a low-lying swamp the southeast wall. Past this morass, at the edge of the current encampment, the ground began to ascend in a gentle east–west running ridge to the point that it dominated the fort at a distance of five hundred to seven hundred yards. To the northwest, there was also low ground, but as one moved directly west of the fort, this gave way to terrain that at three hundred yards sat a good sixteen to eighteen feet above the fort. It was along this side that Eyre expected any attack, the rising ground along the other side being too distant and the intervening swamp too much of an obstacle for the enemy to forward any trench works along this line. The southwest ridge was another matter. Short of entrenching and permanently manning this position, there was no easy way to protect the fort in this direction. In the end, Eyre conceded that a determined enemy would probably gain control of these heights and invest the fort, but he felt confident that if the garrison did its duty, they would have to pay a heavy price in doing so, particularly if they attempted to descend these heights, which would expose their trench works to the fort's guns.[74]

On the morning of September 30, nearly a quarter of the army formed up before the picket fort. As the men stamped their feet and glanced at the low clouds, Captain Eyre pointed at the wooden structure and ordered it torn down. There was a great deal of grumbling as the men went about the task, in part over the spotty showers that hampered the work and in part over the wisdom of building a fort only to tear it down when it was nearly finished. With this accomplished, work began in earnest on what one soldier referred to as "the mountain of sand." The title, although hardly endearing, was not far from the truth. Fort William Henry, like Fort Edward, was essentially an earthen box. Thirty-foot-wide ditches were dug in a rough outline of the fort, which, because of the lay of the land, resulted in a trapezoidal outline instead of the customary square.

Within these trenches, a network of vertical logs reinforced at two-and-a-half-foot intervals by horizontal timbers was erected to form the skeleton of the fort. This framework was then packed with earth in a backbreaking task involving wheelbarrow, shovel and ample amounts of muscle. After ten feet in height, only the forward twelve-foot portion of the wall—or the parapet, as it was referred to—was carried on to a final height of sixteen feet. The rear portion of the wall, or the rampart, formed a platform that allowed troops to move along the length of the walls and on which the cannons were mounted to fire through slots in the parapet. Firing steps were then added to the inner portion of the parapet to allow troops to fire over the wall at an advancing

A plan of Fort William Henry by Captain Eyre. This plan shows the surrounding terrain and the fortified camp on Johnson's former 1755 campsite. The map also shows the advance of the French army during the Battle of Lake George on September 8, 1755. *Library of Congress, Geography and Map Division.*

enemy. At the corners of the fort, bastions were added, protruding sections that allowed the fort's defenders a clear field of fire along the connecting walls. Within these bastions were the fort's magazines and bomb-proof casemates for the troops. To further hinder an attack by storm, sharpened poles were hung off the parapet, and a thirty-foot-wide ditch, bisected by a wooden palisade, was dug about the perimeter of the fort with the exception of the side facing the lake, where, because of the sloping terrain, a ditch was

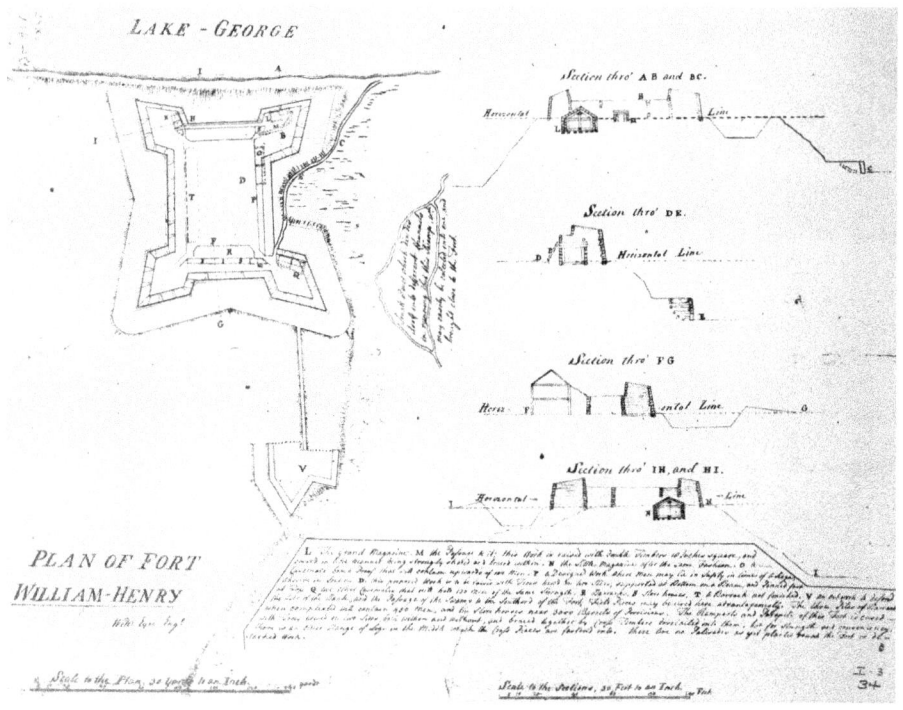

A plan of Fort William Henry by Captain William Eyre showing the positioning of the fort next to the lake and series of cross-sections of the structure. *From Hulbert,* The Crown Collection of Photographs of American Maps, *Series I/3 (1907).*

deemed unnecessary. The dirt from digging the ditch was piled on the outer portion of the ditch to form the fort's glacis, a sloping mound that shielded the fort's walls by limiting the portion of the wall exposed to enemy fire. The final touch was the addition of a ten-foot-wide gate near the southeast bastion and the construction of barracks and storehouses along the curtain walls to house the garrison and their supplies.[75]

The artillery was dragged into Fort William Henry on November 12, and the next morning, the Union Jack was hauled up the newly fashioned flagpole. There was still much to do within the fort, primarily concerning the barracks and some of the firing platforms for the larger guns, but the fort was deemed prepared to accept its garrison. The latter had been agreed upon in a November 18 council of war to consist of 500 men drawn in set proportions from the participating colonies. At the insistence of the various colonial commissioners, however, this was reduced to 430 men a week later. Like Colonel Nathan Whiting at Fort Edward, the new commander of Fort William Henry, Colonel Jonathan Bagley, found his numbers far fewer than promised, and like Whiting, he was promised that the deficiencies, both in men and supplies, would be rectified. Armed with his orders and a memorandum prepared by Captain Eyre on the defense of the fort and the less inspiring topic of honors of war for a surrendering garrison, Bagley assumed command as Johnson led his army back to Albany on November 27.[76]

Both Fort William Henry and Fort Edward would play a prominent role in military operations along the Hudson-Champlain frontier. In command of Fort William Henry, a recently promoted Major Eyre repelled a French winter attack on the fortification in the spring of 1757. While the French war party of 1,500 men managed to burn many of the vessels assembled at the fort over the previous years, they quickly discovered that attacking and seizing the fort without cannons was impossible.

That summer, Fort William Henry's defenses would be put to the test when the Marquis de Montcalm, at the head of seven thousand French and Indian troops, ascended Lake George and invested both the fort and the fortified encampment. Any hope of seizing the British position by surprise was quickly dispelled when the fort's guns began firing warning shots. It was not enough, however, to stop the more numerous French from landing and investing the fort and the nearby British encampment. Once Montcalm was convinced that he had surrounded the British position and cut its communications with Fort Edward, he turned to Captain Desandrouins of the royal engineers to conduct a formal siege of the fort.

After conducting a thorough reconnaissance of the fort and factoring in the current positioning of the army and the terrain before him, Desandrouins had little doubt about how the siege works would proceed. First, the artillery would be landed in a protected ravine behind Artillery Cove, about seven hundred yards from the fort. From there, a main trench would be started in a southerly direction. When the work had progressed far enough, a pair of firing parallels would be dug. The first would be constructed along a branch to the left and the second along a path cut a little farther forward to the right.

At five hundred to six hundred yards from the fort, neither battery was close enough to achieve the desired aim of breaching a wall. Instead, they would cover the continuation of the main sap, which, after negotiating a small marsh, would push forward into the garrison's gardens along the fort's western wall. There a final set of parallels would be constructed at point-blank range, about one hundred yards from the structure, close enough to guarantee a breach in its walls. With the exception of a few obstacles, such as

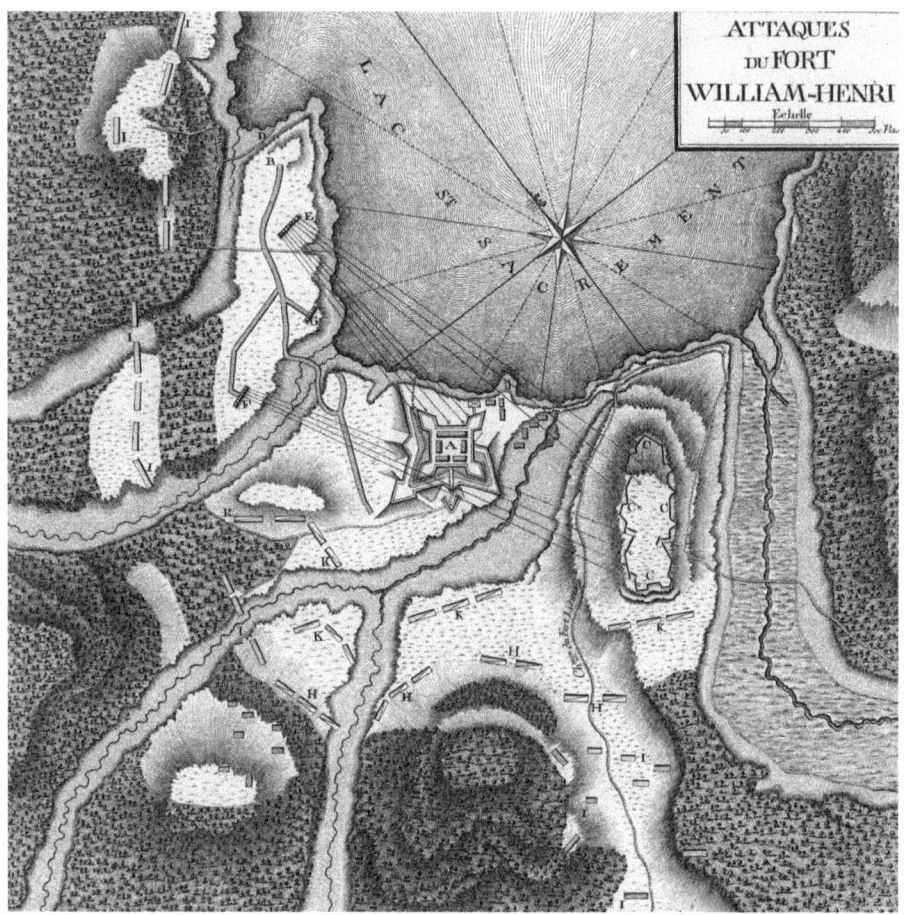

A French map of the siege of Fort William Henry, 1757. The French batteries and siege trenches are shown, as is the position of the French troops encircling the two British positions. *NAC, NMC7805.*

the construction of a dry ditch over a part of swampy ground, the approach proved to be an efficient and successful plan.[77]

In the end, the reasons for the famous British surrender had little to do with the fort. After three days of continuous bombardment the fort's commander, Lieutenant Colonel George Monro, ordered engineer Adam Williamson to make an inspection of the fort. Williamson found that the structure had stood up well to the bombardment. The parapets along the northeast and northwest bastions had the upper three feet blown away, and the casemates had been damaged by exploding shells, but otherwise the fort was intact.[78]

Instead, two other elements were responsible for the defeat. First, the garrison's cannons were in a poor state and continued to fail throughout the siege. Second, with British forces strung out from Albany to Fort Edward, the decision was made not to risk this last post and the New York frontier by marching to the aid of Fort William Henry.

After Monro's surrender, Montcalm destroyed the fort, which was never rebuilt by the British. A new structure named Fort George was laid out on the site of the fortified camp in 1759, but only one bastion was actually completed. This fortified post would be active throughout the remainder of the French and Indian War and was still maintained by a handful of troops during the American Revolution to facilitate communications to and from Fort Ticonderoga and Crown Point.

While Fort William Henry would stand for less than two years, Fort Edward would become the lynchpin of the British defenses on the New York frontier. The stronghold would be involved in every major British campaign on the lakes from 1755 to 1760, and during this time, work was continually being done on the fort and its outworks. At the height of the Fort William Henry crisis in early August 1757, chief engineer Colonel James Montresor reported on the status of the fort. Four casemates had been constructed, one of which was designated as the fort's magazine, and final touches were being put on the officers' barracks, the well and the fort's oven. The ditch about the fort, the ravelin protecting the north gate, the northwest and northeast bastions and the curtain wall running between them were all finished, with the exception of a few minor parapet issues. The southeast bastion was also finished, as was the curtain wall between it and the southwest or water bastion, although the parapet along this wall still needed work. A fascine and gabion parapet had been put in place between the southwest bastion and northwest bastion, and across the river, a fortified camp had been built on Rogers Island.

In addition to this, the fort no longer lacked firepower. Thirty-nine cannons now lined the walls, the heaviest of which were eight eighteen-pounders, six nine-pounders and four large mortars. It was a far cry from the early days when a few six-pounders were considered a welcome gift. To back these cannons, the fort was now capable of taking a huge garrison. Montresor estimated provisions for 1,434 men within the fort itself, 731 in the fortified camp across the river on Rogers Island and another 1,770 employed in makeshift lines before the structure, making for a grand total of nearly 4,000 troops.[79]

With the capture of Forts Carillon and St. Frederic in 1759 and the diminished threat to the posts on the New York frontier, Fort Edward's role

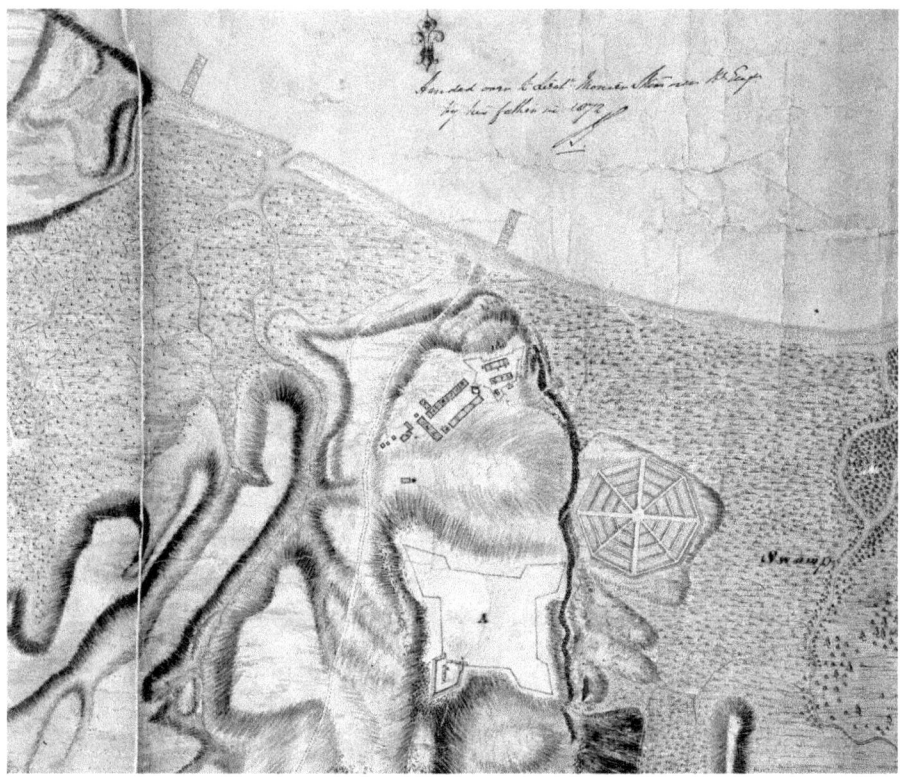

A 1759 plan showing the layout of Fort George (A). Only the southwest bastion would ever be finished, and this was later modified into a stand-alone stone redoubt. The ruins of Fort William Henry can be seen in the upper left-hand corner of the image. *Boston Public Library, Norman B. Leventhal Map Center.*

was reduced to more of that of a staging and supply facility. The fort would remain in operation after the war, but as the years progressed, the garrison was reduced to a handful of men and the structure decayed to the point that by the time of the American Revolution, it could only be considered a fortification on paper.

Chapter 4

THE LOWER HUDSON VALLEY

At the south end of the Hudson waterway is one of the first forts built in North America, Fort Amsterdam. Erected in 1625 to protect the fledgling Dutch colony on Manhattan Island, the fort was built like Fort Orange—an earthen structure within a wooden framework. As would be expected, Fort Amsterdam suffered from the same design problems as Fort Orange. The wooden portions of the fort soon rotted away in the wet climate, and the earth that was contained within these confines seeped out.

Less than twenty years later one witness wrote, "The Fort is defenseless and entirely out of order," while another noted:

> *The Fort under which people will take shelter, and from which, it seems, all authority proceeds, lies like a mole-hill or a ruin. It does not contain a single gun-carriage, and there is not a piece of cannon on a suitable frame, or on a sound platform. It was proclaimed, at first, that it should be repaired, constructed with five bastions, and be made a first class fort. The Select men were also asked for money for the purpose; but they excused themselves on the plea that the people were very poor.*[80]

The following year it was recommended that "it would be advisable immediately to construct Fort Amsterdam of stone; for it is now in such ruin that men pass into it, over the walls, without making use of the gate." Most would not question the advice, but as before, it was really more a question of

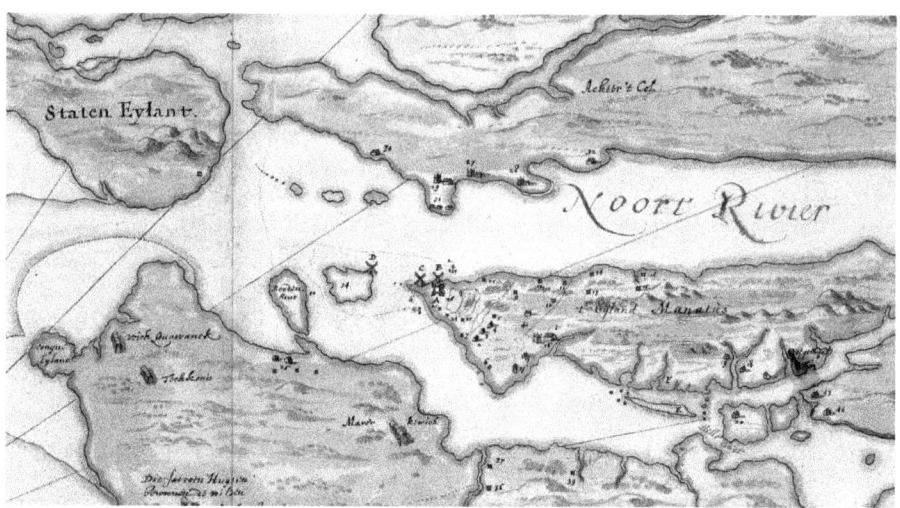

A portion of a 1639 Dutch map showing Fort Amsterdam, Manhattan and the North or Hudson River. *Library of Congress, Geography and Map Division.*

money than direction. As a temporary fix, it was agreed to try and prevent the decay of the fort's earth walls by covering them with sod, but "as the soil is sandy and the foundation weak, the sods mostly sagged and fell to pieces." The inhabitants' swine contributed to the problem as well, "whereupon the Director ordered the soldiers to shoot whatever hogs came there." Nor were the problems confined to the fort itself, as one Dutch officer pointed out. "The gun-carriages, and whatever belongs to them, are bad and mostly all unserviceable; if new ones be made, the sun and heat will immediately dry them up, and cause them to split and warp, unless tarred, they will be, in a short time, rendered wholly useless."[81]

Fort Amsterdam would serve as a refuge for Dutch settlers during the brief but tumultuous history of the colony. A series of conflicts with local tribes, starting with Kieft's War and extending through the Second Esopus War, would on more than one occasion force large portions of the population to seek shelter under the fort's guns. Although the structure was in disrepair, armed with cannons and backed by the now concentrated militia in the town, it would extract a high price on any enemy attempting to take it by storm, and as such, it was never truly challenged during these conflicts.

Beyond neglect and the ravages of the seasons, the first real test for Fort Amsterdam came with the start of three Anglo-Dutch wars in Europe. In 1654, near the end of the First Anglo-Dutch War, a paper circulated in England detailing a series of imaginary Dutch atrocities and a sinister plan

The Lower Hudson Valley

The last governor general of the New Netherlands, Pieter Stuyvesant. For almost twenty years, Stuyvesant guided the troubled colony through economic calamities, Indian conflicts and threats from colonial neighbors. *From Abbott, John S.C.,* Peter Stuyvesant, the Last Dutch Governor of New Amsterdam *(1898).*

to inspire an Indian uprising against the English inhabitants of the disputed Connecticut River Valley. The paper, along with a number of colonial letters and the urgings of well-placed individuals, was enough to convince Oliver Cromwell that military action was required. The Protector dispatched four frigates and 200 troops under the command of Captain John Leverett and Major Robert Sedgwick to assist the New Englanders in their cause. The fleet reached Boston in June, when it found a populace more than willing to assist in the conquest of the New Netherlands. The muster drums were beat and corner criers called upon the men of New England to join the crusade against the Dutch menace. By mid-July, the efforts had netted 633 volunteers, including a troop of horse, which, along with the necessary supplies, were loaded onto the English warships.[82]

The governor of the New Netherlands, Peter Stuyvesant, knew of the English fleet, but he was ill prepared for its arrival. A palisade and shallow ditch had been incorporated into the defenses of New Amsterdam a few years before, when news of the war first reached the colony, but since that time little else had been done. Work was restarting on repairing Fort Amsterdam, but it had to be done secretly so as not to alarm the inhabitants, most of whom, it was known, would not fight against such overwhelming odds. With only one hundred soldiers to man a crumbling, weather-stripped fort and a militia unlikely to answer a call to arms, New Amsterdam's prospects looked bleak.

But history was not finished with the New Netherlands. As the English fleet was making its final preparations, a London merchant ship arrived at Boston with news that the war was over. A day later, the news reached New Amsterdam. Much to the relief of the residents and their governor, the colony had been spared.

The reprieve would last only ten years. With the Dutch and English once again at war, in 1664, Charles II, responding to New England claims and complaints regarding their Dutch neighbors, dispatched four warships and 450 troops under the command of Major Richard Nicolls with orders to seize the New Netherlands. Nicolls and his flagship *Guinea* arrived in Boston in late July. The remainder of his little fleet, scattered by a storm, entered the port over the next few weeks. The vessels were resupplied and volunteers recruited for the expedition. With all the preparations completed, the *Guinea* set sail for New Amsterdam and dropped anchor near Coney Island on August 26, 1664. A few days later, Nicholls was joined by the rest of the fleet, which consisted of three frigates, a brig and 300 troops.[83]

Stuyvesant was at Fort Orange when news of the English squadron's departure from Boston reached him. He rushed back to New Amsterdam just in time to witness the arrival of the *Guinea*. On August 29, three other warships dropped anchor in the inner harbor, which left no doubt as to English intentions. The governor began organizing a hasty defense. Every third man within the limits of New Amsterdam was ordered to begin work on the city's defenses with spade, shovel or wheelbarrow. Carriages were hurriedly constructed for eight cannons that were to be added to the fort's current complement of fourteen. On the land side of the island, work was started on a trench and a rude breastwork. Troops stationed at Esopus and other outlying posts were recalled, and Fort Orange was summoned to send whatever help it could muster. But no reinforcements were forthcoming. These communities sympathized with the governor's plight, but being more afraid of the natives than the English, they refused to release their soldiers.[84]

While Stuyvesant prepared New Amsterdam, Nicolls seized Staten Island and blockaded Manhattan, cutting off all communications between the island and the rest of the colony. On September 2, Nicolls officially summoned Stuyvesant to surrender Fort Amsterdam and the island of Manhattan. Stuyvesant stalled for time while the defenses of New Amsterdam were readied and the militia called out. The Dutch governor sent the summons back, noting that it was not signed. Nicolls, with a sly smile, signed the document and penned a short apology before having both

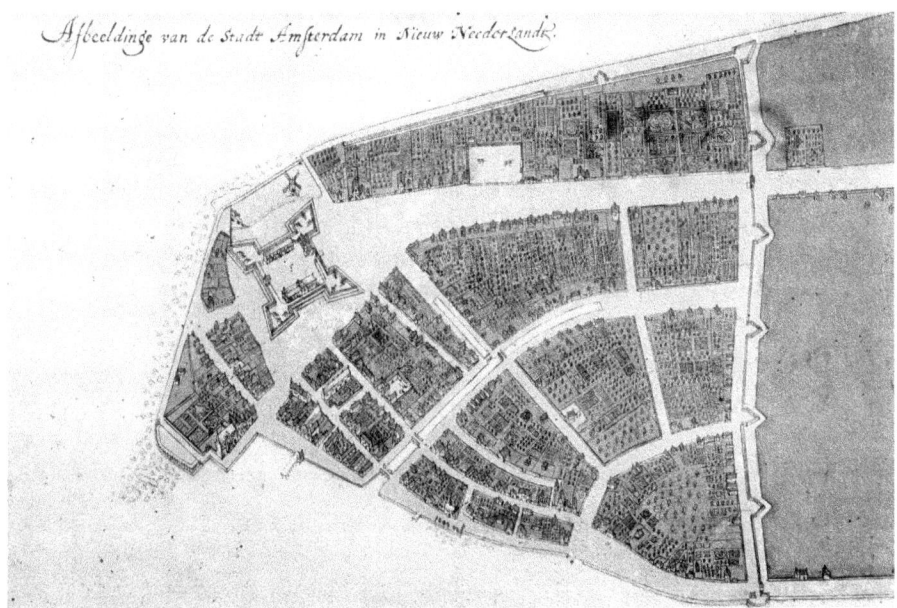

A plan of Manhattan in 1660 showing the layout of Fort Amsterdam and the palisade works at the edge of the town. *New York Public Library.*

delivered the next day. Stuyvesant penned a long reply to Nicolls's summons in which he outlined the territorial rights of the New Netherlands, their just claim to the territories they occupied and the tenuous if not outright false assertions made by the English to these lands. He then placed this letter in the hands of four trusted envoys and directed them to personally argue the matter with Nicolls.[85]

The Englishman, however, was not interested in arguing. Searching for an opening, the envoys replied, "Friends will be welcome if they come in a friendly manner." Nicolls, irritated with the delays, responded sarcastically, "I shall come with my ships and soldiers, and he will be a bold messenger, indeed, who shall then dare to come on board and solicit terms." When the dejected Dutch deputies threw up their arms and asked, "What, then, is to be done?" Nicolls ended the meeting with, "Hoist the white flag of peace at the fort, and then I may take something into consideration."[86]

The brief parley gave Stuyvesant time to examine his situation. Fort Amsterdam was a dilapidated structure incapable of surviving more than a few cannon shots before collapsing, and mounting only a few light brass cannons, it was doubtful whether it could even damage the English warships, which bore tenfold as many guns of much larger caliber. Not that it would matter. The master gunner had estimated that there was only enough powder for a day's firing. The fort's 150 company soldiers could be counted on to fight, but the 250 or so militia of New Amsterdam had refused to become involved in such a one-sided affair. All work on the city's defenses had stopped, and not a single burgher would answer the call to arms. Still, Stuyvesant was determined to resist. A longtime soldier, he was loath to give up without a fight. The cannons were loaded, and the smell of lit matches drifted on the fine autumn breeze as several leading citizens pleaded with the governor not to undertake such a folly.

Perceiving that Stuyvesant was determined to resist, Nicolls began landing troops below Breukelen and ordered a pair of frigates to take up firing positions before the fort. Surveying the warships from a vantage point in one of the angles of the fort, the governor finally bowed to the hopelessness of the situation and signaled for the Dutch flag to be replaced with a white one.[87]

Fort Amsterdam was now Fort James, named after the king's brother, who had been granted the entire colony that would be renamed New York. Nicholls, who was made governor of the new English colony, repaired the fort with his limited resources. He supplied it with more cannons and its walls were bolstered, but given the state of the structure, it was doubtful whether it would buy the fort more than a few hours if besieged.

With the opening of a third Anglo-Dutch War in 1672, the new governor of New York, Francis Lovelace, attempted to address concerns regarding the defenses of the colony. In July, the militia rolls were filled out and trivial efforts were made to repair the colony's forts. Tensions rose when a pair of Dutch fleets under Admirals Cornelis Evertsen and Jacob Binckes combined to attack an English tobacco convoy in Chesapeake Bay.

Emboldened by their success, Evertsen and Binckes set their sights higher. They questioned a captured English captain on the state of Fort James's defenses. The defiant Englishman claimed that five thousand men stood ready to defend the 150-gun fort. Not amused, the two Dutch admirals questioned a more cooperative subject, who informed them that the numbers were closer to eighty men and thirty guns. Upon hearing the news, the Dutch commanders set sail as the cry "New York, New York!" went out through the fleet.[88]

Evertsen and Binckes entered the narrows of New York Harbor on July 28, 1673, and found little to oppose their progress. Their fleet consisted of twenty-three vessels, of which eight were warships. Governor Lovelace was in Connecticut at this important juncture, so command fell to Captain

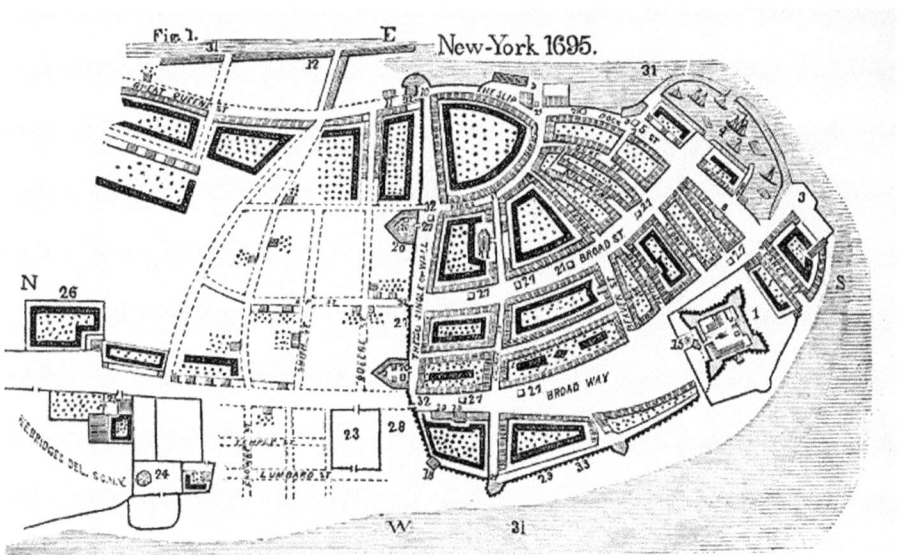

New York City in 1695. The map shows the tip of the isle of Manhattan, which overlooks the Upper Bay. The East River is at the top of the map and the Hudson River at the bottom. *From Miller,* Description of the City and Province of New York, *1695.*

John Manning, a former commandant of Fort Albany. Upon seeing the Dutch fleet, Manning ordered the militia called out and the defenses of the town readied, but like Stuyvesant a decade earlier, no volunteers came to the beat of the drums. To the contrary, the Dutch inhabitants, seeing their countrymen riding at anchor a few hundred yards away, began making threatening speeches and arming themselves.

With a garrison of eighty men and eighteen operational cannons to oppose the fleet before him and a hostile populace at his back, Manning was doomed. Of course, he sent couriers to Governor Lovelace and requested reinforcements from the local garrisons, but neither effort would yield any immediate results. With few options, Manning requested a parley. The Dutch commanders agreed and used the opportunity to present their summons for the fort's surrender. Manning stalled. When the allotted time passed, the Dutch began landing troops on the west bank of Manhattan Island and moved their warships to within musket shot of the fort. Another half hour passed before the volley of one hundred cannons shattered the still July day. The fort responded with a volley of its own, which struck the Dutch flagship, but in doing so, most of the stronghold's cannons leapt off their decrepit mounts and could not be fired again.

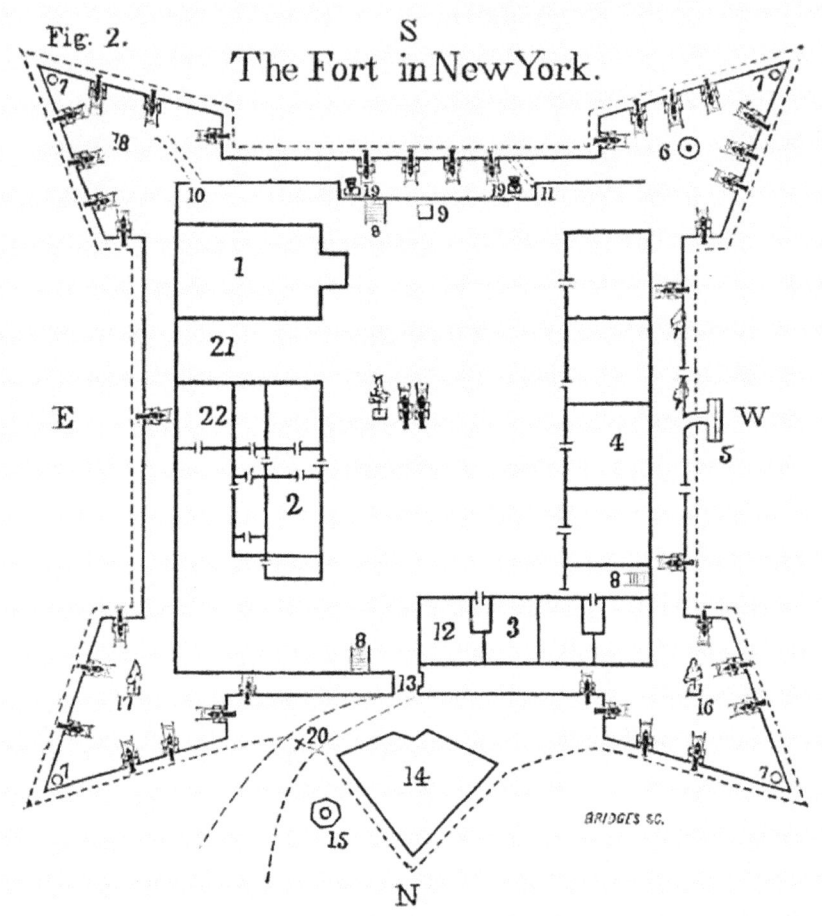

THE EXPLANATION OF FIG. 2.

1. The chappell.
2. The governor's house.
3. The officers' lodgings.
4. The soldiers' lodgings.
5. The necessary house.
6. The flag-staff and mount.
7. 7. The centry boxes.
8. 8. Ladders to mount the walls.
9. The well in the fort.
10. The magazine.
11. The sallyport.
12. The secretary's office.
13. The fort gate.
14. A horn-work before it.
15. The fort well and pump.
16. Stone mount.
17. The Iron mount.
18. The Town mount.
19. 19. Two mortar pieces.
20. A turn-stile.
21. Ground for additional building to the governor's house
22. The armory over the governor's kitchen.

Fort William Henry (Fort James), 1695. Although a great deal of work was put into the structure during the early years of King William's War, it was still no match for a concerted naval attack. Fortunately, none occurred. *From Miller,* Description of the City and Province of New York, *1695.*

The Dutch fired several more broadsides, which killed a handful of the fort's garrison and silenced the few cannons that were still serviceable. With over six hundred Dutch troops approaching from the land side of the fort and his defenses on the sea side shattered, Manning beat a parley to ask for terms. Somewhere along the line, his orders were misinterpreted and the colors were struck. It was a mere detail. Within a few days, the remaining forts in the colony had surrendered, and New York was once again Dutch.[89]

The Dutch return was to be short-lived. The next year, New York was returned to English rule as part of the treaty that ended the Third Anglo-Dutch War. A new governor, Sir Edmund Andros, would take command of the colony in October 1674, and like his predecessors, Andros would repair Fort James, stock the stronghold with hundreds of muskets and rearm it with forty-six cannons. Lack of funds, however, would limit his and future governors' efforts.

Real fear for the safety of the fort, and the colony in general, arose when news of war between England and France reached the colonies in the summer of 1689. There was instant apprehension along the Eastern Seaboard. A major French naval expedition was likely to overwhelm any port in the colonies, even Fort James, which guarded New York City. Although frequently repaired and cited for its deficiencies, Fort James was still a crumbling mess. One witness who surveyed the fort in June 1689 wrote that the structure was "much out of repair most of the great gunns not fit for servis: very few platforms for gunns to play: and by the account showed us taken by skillfull honest men of the powder that of 50 barells: but one good and a considerable part not fit for any servis, and the rest would not sling a bullet half over the River."[90]

By July, repairs to the curtain walls and bastions of the newly named Fort William Henry were underway, as was work to restore the interior buildings and the wooden palisade that encompassed the landward side of the town. To try and overcome the fort's shortcomings, a number of local gun batteries had been erected, with two of the primary ones along the east shore of Manhattan at the Slip (Burgher's) and at Whitehall. These were complemented by temporary emplacements along the banks of the Hudson River on the west side of the island. On paper, it looked impressive. On a good day, some fifty guns could be brought to bear against an enemy vessel entering the upper harbor, but the truth was that good days were long behind the cannons that manned the city's defenses. Neglect had made half the guns and their carriages unserviceable and the rest a frightening proposition for the crews that had to fire them. Soldiers,

A painting of Fort George, circa 1740. *Albany Institute of History and Art.*

cannons, shot and powder were requested from England, as were a pair of Royal Navy frigates to defend the coast, but for the time being, the troops available to Governor Jacob Leisler—perhaps six hundred if the militia is included—were left to their own devices in a town that had already been captured twice in its short history.[91]

In 1701, Colonel Wolfgang Romer surveyed the harbor and its defenses. His recommendations called for a battery of thirty guns on both sides of the narrows formed by Staten and Long Islands, a blockhouse and a fort of fifty guns on Sandy Hook to force an enemy fleet to stand farther out to sea during operations. He also recommended a battery of a dozen guns at the narrowest part of Hell Gate to prevent access to the East River via Long Island Sound. The sensible recommendations were at first heeded, and money was allocated for the proposed batteries at the Narrows in 1703. But the defenses never materialized, the money instead being channeled into the construction of a new mansion for Governor Cornbury. In 1711, Governor Hunter pressed the New Assembly with the urgency of carrying through with this project, but remarkably, it would not be until the late 1750s that defensive recommendations for this area were actually put into place.

The fort, renamed Fort Anne in 1703 and finally Fort George in 1714, remained a controversial centerpiece of the city's defenses. Renovated half

The Lower Hudson Valley

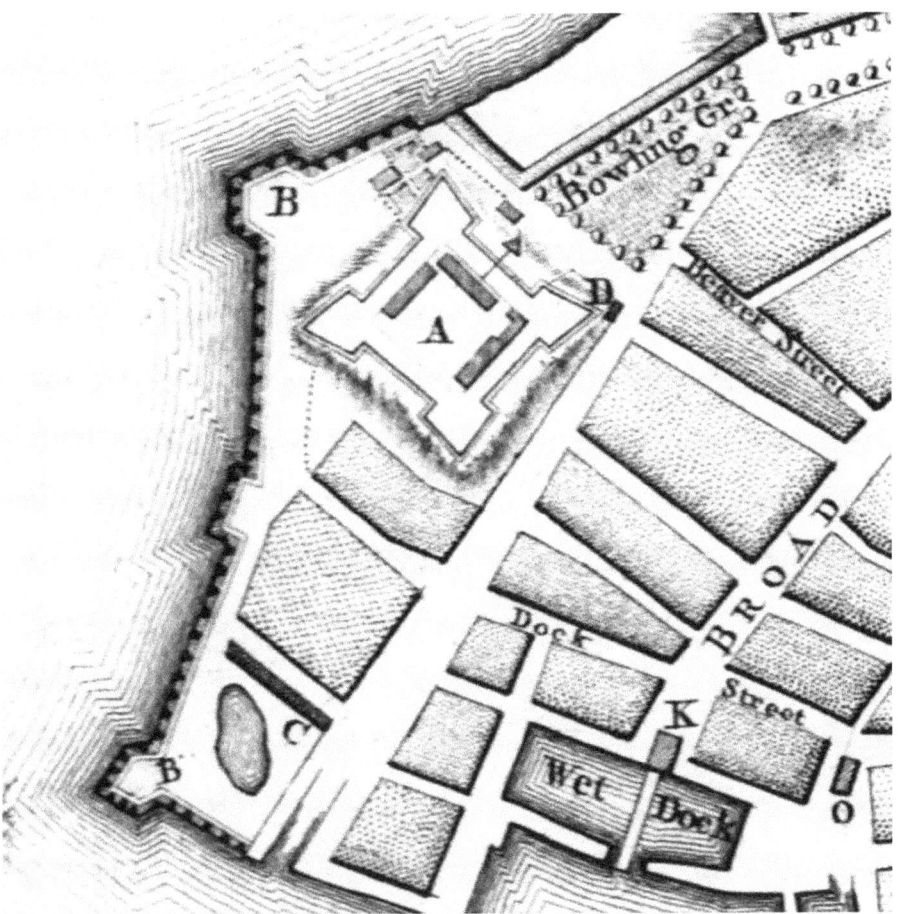

A portion of a 1775 map of New York City by royal engineer Captain John Montresor. Fort George (A) and its supporting gun batteries (B) are detailed in this section. (C) is the military hospital, and (D) is the general secretary's office. *Boston Public Library, Norman B. Leventhal Map Center.*

a dozen times since its capture from the Dutch, by the year 1738, the structure had reached a point that the governor of New York complained of its condition and the lack of military supplies within its walls. Fortunately, the fort was backed by a nearby battery of fifty guns erected in 1735 to control access into the upper harbor. As for Fort George itself, the interior buildings were renovated after a fire in March 1741 destroyed most of them, but from this point on, the structure, mounting only a handful of twelve- and nine-pound cannons, took on a secondary importance to the nearby battery.[92]

Fort George and New York City would not be challenged in King George's War or during the French and Indian War, which would follow a few years later. By this point, the population of the town, combined with the gun batteries positioned along the shores of the island and along the Hudson and the increased presence of the British navy, made any thought of a naval expedition against the town a remote possibility.[93]

Appendix A

GLOSSARY*

BASTION: A small blockhouse structure, typically diamond in shape, located at the junctions of the fort's walls. While platforms on the bastions were used for the fort's guns, the primary purpose of the structure was to provide flanking fire against any troops attempting to storm the adjacent curtain wall.

BATEAU: A small wooden shallow draft boat, similar to a whale boat, which could either be rowed or powered under a simple square sail arrangement. In some cases, these vessels would also be armed with small cannons on their bow and/or stern.

BOMB: An exploding mortar round.

BOMB-PROOF: A shelter with overhead cover built to withstand the impact of an exploding mortar round.

BREASTWORK: A field fortification made of earth and timber.

CANNON: The cannon of this period was either made of brass or iron and fired solid ball projectiles ranging from three pounds to forty-eight pounds. While the heavier guns were typically employed in European-style fortresses or ships of the line, along the Old Invasion Route, the largest cannons used were a pair of thirty-two-pound guns mounted in Fort William Henry. Smaller twenty-four-pound iron cannons, which still weighed a ton and a half, were the popular siege gun employed along the waterway.

---

*. For additional terms and definitions, see Sabastien Vauban, *A Manual of Siegecraft and Fortification* (Ann Arbor: University of Michigan Press, 1968).

Appendix A

CASEMATE: A bomb-proof shelter designed to either protect the garrison or the fort's magazine from mortar fire.

CHEVAL DE FRISE: This defensive structure is a row of sharpened stakes driven into the ground and angled toward the attacker. Often, trees were used in place of stakes, with their branches sharpened to points. This field fortification was originally used to repel cavalry charges, but it was also used to slow an infantry attack.

COVERED WAY: This is a platform cut into the ditch side of the glacis to allow for troop movement and protect small arms fire.

CURTAIN WALL: An interconnecting wall between two bastions.

DEMILUNE: A smaller crescent-shaped fortification typically placed in front of a curtain wall. Such structures are small forts unto themselves and are frequently connected to the nearby wall via a drawbridge.

DRY DITCH: A technique used to cross a swamp or marshy land when a traditional trench is impractical. In such cases, a causeway is built and then shielded with gabions and a timber roof.

EMBRASURES: Openings cut into the parapet to allow for either musket or cannon fire. (See *merlons*.)

FASCINE: A bundle of small branches or used to support earthworks and in particular to brace the walls of a siege trench when crossing wet ground or to fill in a ditch to allow for crossing.

GABION: A barrel filled with earth. Gabions were typically used as temporary field fortifications or as temporary repairs to damaged or incomplete fortifications.

GLACIS: Sloping ground placed in front of the fort's walls to reduce the amount of wall that can be seen and targeted by enemy cannon fire. Shots that are too short deflect off the glacis and over the fort's wall.

GRAPESHOT: Bags of musket balls fired out of cannons. This type of shot was used to repel an infantry attack.

MACHICOLATION: Extended portions of a wall or tower designed to allow the defender to shoot down or drop objects on attackers at the base of the supporting wall.

MERLONS: The solid portion of the parapet. These may be pieced with slit embrasures to allow for covered musket fire, or they may be located on either side of a larger embrasure, which allows for cannon fire.

MORTAR: An artillery piece designed to fire exploding rounds (bombs) in an arching trajectory so they will clear the fort's walls.

PALISADE: A wooden wall typically made of trees placed vertically in a filled ditch and then bound together.

Appendix A

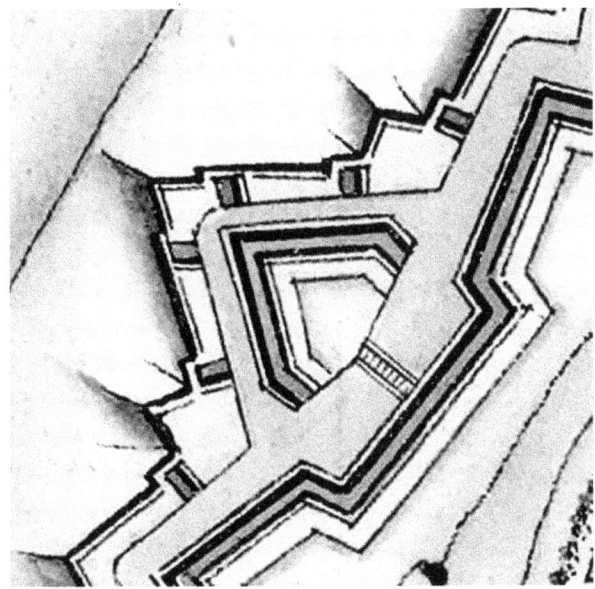

Left: A portion of a 1735 French plan of Fort Louis showing one of the many demilunes surrounding the fort. Note the drawbridge that provides access to this structure from one of the fort's walls. The use of demilunes made it much harder to target large portions of the fort's walls. *Bibliotheque Nationale de France.*

Below: Cross-section of a fort's wall and outer works. *Author's collection.*

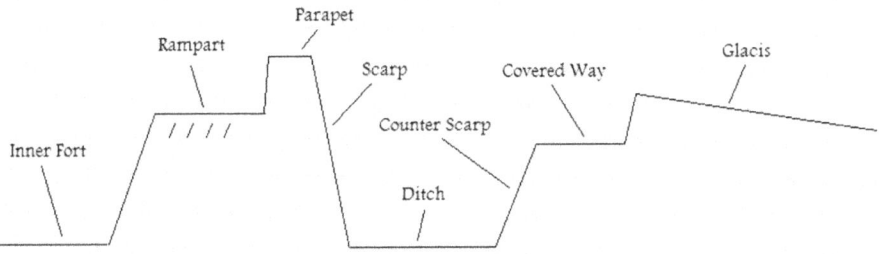

Cross Section of a Fort's Outer Works

PARALLEL: A perpendicular, or near perpendicular, extension of the main sap designed to employ a battery of guns from which the besieged fort can be bombarded.

PLACE OF ARMS (OR PARADE GROUND): The open interior portion of the fort where the garrison can be assembled.

RAVELIN: A redan.

REDAN: A *V*-shaped fortification where the point is aligned with the direction of an enemy attack. Redans were typically employed as outworks or field fortifications and were often incorporated into fort walls to provide flanking fire down the length of the wall.

REDOUBT: A self-contained fortification with a ditch about its perimeter and earth and timber walls.

Appendix A

SAP: A trench.

STORM (TO TAKE BY): Meaning an infantry attack on the fortifications. Such measures were either combined with the element of surprise or viewed as last resorts given the chances of success and the large number of casualties that would come from the effort.

TOISE: An old French unit of length being equivalent to 6.39 English feet.

Appendix B

DOCUMENTS

REPORT ON THE FRONTIER DEFENSES OF NEW YORK

By Colonel Wolfgang Romer
August 26, 1698
(*NY Col. Doc.* 4, 440–41)

In obedience to Your Excellency's orders, I proceeded, on the 18th of May of the present year, 1698, to the frontiers of the New-York government, and, in the first instance, towards Albany, Schanegtade [Schenectady], Kanestigionne [Canastigione], and the Half Moon; and, after having observed these places, I found the city of Albany, situate on the Hudson river 144 miles north of York; an important frontier, as well as Schanegtade, 20 miles West of Albany, on the Great Mohawk river; but these frontiers are neglected, built of wood and palisades of poor defense. Saving better judgment, my opinion would be to build stone forts there, constructed and proportioned according to the respective situations, and the importance of the one and the other of these two places. For I consider if these two places should one day fall into the hands of the enemy, the provinces of York, Jarsé, Pensilvania and Connecticut, would be obliged, in a short time, to submit; and that Maryland, Virginia and New England would, consequently, greatly suffer. Also, as York is the depot of all the Islands for flour, grain and other provisions, these

Appendix B

would experience a very serious injury. In regard to the other places, Kanestigionne and Half Moon—the first; 12 miles north-west of Albany, on the Great Mohawk river, and the other, 12 miles north on the Great Hudson river—they are to be regarded as the barriers of the two abovenamed frontiers on the side of New France; and, therefore, of great use in time of war for the preservation of good communication between the two principal frontier posts which I have already mentioned; so much so, that I should deem it proper to build a good guard house, or stone redoubt, for the accommodation, in case of necessity, of 30 or 40 men; and in case of war, a strong, well flanked palisade could be attached to it, to serve as a retreat for the neighboring settlers who reside 1, 2, 3 and 4 miles distant, one from the other.

Regarding Cheragtoge [Saratoga], a post on the Hudson river, 28 miles north of Half Moon, I could not get there, though I had set out for that purpose, in consequence of the freshet in the rivers and other impediments, which it was impossible for me to surmount. I shall observe, however, with submission to Your Excellency, that I learned, by minute inquiries I instituted, that the farms, which were only seven in number, as well as the fort which was built there in Leisler's time, have been entirely ruined by the last war; since which time they have never been thought of, and the settlers have never thought of returning thither; and, also, because the French claim this country as dependent on them, notwithstanding we have had possession of it a great many years. I think it would not be useless to have a small fort built there of palisades, with a small stone tower in the centre, to maintain possession, and encourage the settlers to build and take up their residence there again. In time, the land can be cleared and the timber cut down, in order to open the country, render it accessible, and to establish an easy communication, so as to support said fort and the settlers, in case of need. Otherwise, a garrison in that place would be, as it were, abandoned.

As for the other reflections to be made on the frontiers, of which I have just spoken, Your Excellency can see them in the plans I now, with submission and respect, lay before you.

W. ROMER.

APPENDIX B

REPORT ON THE HARBOR OF NEW YORK

By Colonel Wolfgang Romer
New York, January 13, 1701
(*NY Col. Doc.* 4, 836–37)

My Lord,

Pursuant to your Excellency's verbal order of the 7th of December 1700, to measure the distance across the Narrows, and to sound the depth of the water there, as well as in a second arm of Hudson's River called the Coll [Kill Van Kull], between Staten Island and East Jersey, and to ascertain whether any ships and bomb ketches could come around by Amboy and consequently attack the city of N. York: item, to select a couple of places both at the Narrows and the Coll, where suitable fortifications could be erected, and the enemy thereby be forestalled in his undertakings, I on measuring the same, have found the distance between the heights [*hoofden*] to be one and ½ mile English in breadth from shore to shore.

In regard to the depth of water, I find across from Long Island to Staten island 4. 4. 4. 4½: 6. 11. 12. 13. 12. 9. 6. 6. and 5 fathoms right under the shore of the aforesaid Staten island. By the second sounding from Staten Island to Long Island ¾ of a mile farther south, where the river is narrowest, I find right under the shore, 5. 6. 12. 14 and 15 fathoms in the deepest part of the channel; this depth then falls off immediately to 6. 2 and 1½ fathom of water where there is a Bar which, with a point northerly towards N. York, runs into Long island and westerly 1/6 part across the Narrows, and S.S.E. towards Sandy Hook runs past Long Island hook where it shoots around E and E by North.

Now for the fortification of the Narrows, I am of opinion that there ought to be, both on Long, and Staten Island, a sufficient Battery with a good Redoubt on each height, enclosed with proper lines of defense communicating with the respective Batteries, and that each be furnished with 30 guns carrying 18 to 24 lbs. ball.

In regard to the other branch of the Hudson River, called the Coll, between Staten Island and East Jersey, I have sounded it from Amboy up to Tampsons [Thompson's] point and Elizabeth town and find from Amboy to the abovenamed points 8. 7. 6. 5. and 4 fathoms of water, it then becomes shallow with a very crooked Channel +0 having no more than 11 to 12 feet of water at spring tide, so that a ship can indeed come up as far as

Appendix B

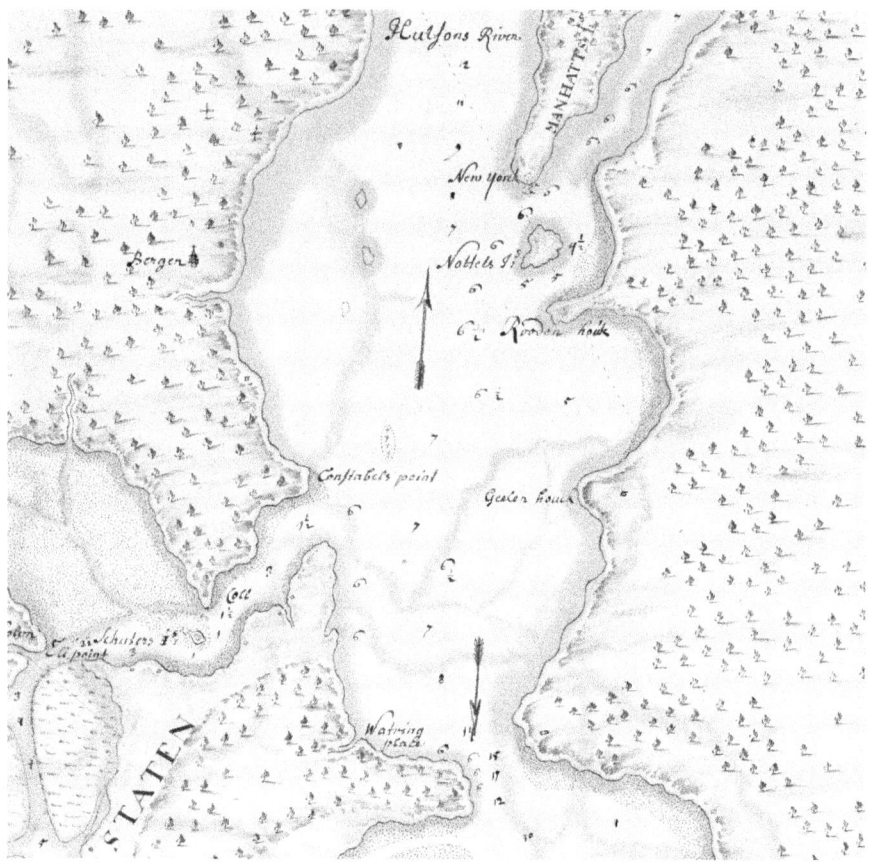

A portion of Colonel Romer's survey of New York Harbor, 1700. *Boston Public Library, Norman B. Leventhal Map Center.*

Tampson's point aforesaid, but with difficulty, because the river runs narrow and crooked. In order, now, to hinder the approach of any vessel, I am of opinion that it can be effected by the erection of a battery on Schutter's island 12 to 13 miles from New York; with this, it is impossible for any ship, sloop or boat to run up or down.

I consider myself bound particularly to submit to your Excellcy the great importance of Sandy Hook, and entertain that opinion, because reason and the Rules of War agree, that an enemy must always be kept as far off as can possibly be done, that a good blockhouse and other fortification ought to be erected on the aforesaid Hook, as they would be very useful there, the channel and entrance being very narrow, and vessels on that account must pass immediately under this Hook, whilst the East banks lie sheer by and

Appendix B

over the Hook running up to the North and East, and it is therefore very dangerous. For these reasons a good Blockhouse and Fort of 50 guns might answer, and prevent any enemy coming by water into my bosom, and oblige him to stand out to sea on a dangerous coast.

Further and lastly, an enclosed battery of 12 to 13 guns ought to be erected at the narrowest part of Hellgate, to prevent the entrance of an enemy at that point also.

All this being done, I am persuaded an enemy will bethink himself a hundred times before he will meditate any attack on New York.

W.W. ROMER

Directions to the Commandant at Fort William Henry

By Captain William Eyre
George Camp, November 25, 1755
(*Johnson Papers*, vol. 2, 328–32)

Directions to be observed, and followed after, as much as Circumstances Will admit in Fort William Henry in case of an Attack by Artillery, Upon Notice of the Enemy's approach, the Commanding officer is to level every sort of Cover round the Garrison (if not done before) as Soon and as much as possible his time Will allow, and to take every Method to deter if not hinder them from getting possession of the Eminence to the South West of the Fort, by keeping a Constant fire of Artillery upon them Should the approach it from the North East, by the West Side of the Lake, as they must be much exposed from the Fort in drawing their Heavy Guns that way: this Method to be Observed on the Supposition that the Lake is not froze, and that the Enemy Will come by Water within near Gunshot of the Garrison before they Land their Force and Artillery. If they should attempt a Siege when the Lake is lock'd in with Ice, they Will be under the Necessity to mount up the Bank on one Side or the Other, for the Surface of the Water is so much below the Garrison that they will not be able to do any Mischief with Batteries on the Ice, besides their being so much exposed, therefore if they march to the Westward of the Lake the Method before mentioned should be observed.

Appendix B

If the Enemy should think it advisable to bring their Artillery to the South, & South East Side, by the East Side of the Lake, or on the Lake, they Will still be exposed coming that Way to the Cannon from the Fort, tho' more remote than the Other, and after they have raised Batteries on their Side, as near as the Swamp Will allow them, yet they Will not, it's apprehended be able to make from those Places practicable Breaches prudent to assault what is chiefly to be feared from Cannon at this distance, is their dismounting of some of the artillery if care is not taken, tho' let what will be done, accidents may, and will happen.

But then if the Enemy, as is very likely will endeavor to Cross the Swamp to the South East of the Garrison, in order to seize the above Mentioned rising Ground to the South West; in that Case, if this be done within Cannon Shot of the Fort it will prove to them a difficult undertaking, besides their Loss, before they can accomplish it; and if this Passage is made further off, they Will find it not an Easy Matter to ascend the very high & Steep Bank, that is to be met with there. However in the End it may be Supposed all those difficulties are to be Surmounted, at the Expenses of Men, & time, and that they get entrenched on this rising Ground, before which is done, the Cannon should be placed as fast & as quick as is possible; and great care should be taken to Secure all those which cannot be made Use of in the most Safe Places which the Commanding officer must be the best Judge of.

When the Enemy get themselves safely Secured by means of trenches and Breast Works, and have or, are rising Batteries, the Mortars as well as the Guns should be at Work, to retard and hinder the Progress of their approaches; when the Enemy begins to descend this Hill, then they become much more exposed, and their approach more hazardous and difficult, if the Garrison Will take their advantage and are obstinate. I may naturally Suppose by the time all the Barracks may be much damaged if not wholly destroyed, by means of Shells, fire, & Shot, but this must be expected, and the Men off Duty to lie in the Casemates where they can repose themselves without Danger; Pains ought to be used to prevent the firing from spreading as much as possible, otherwise, one do not know but it may be possible an accident may happen to one of the Magazines; the Powder should be divided between them. All that can and will contribute to make a Noble Stand, is, by not being intimidated by accidents, considering Maturely the advantages the Works and their design, and being resolute, if it must go, to make them dearly Pay for it, both in loss of time as well as in Blood, should the resolute defense not give time to the Country to

Appendix B

come to its relief; which must certainly happen, if the Garrison Will act on those honorable Terms, and the following the aforesaid Rules, as nearly as Circumstances, and time Will allow.

When the Enemy advances close to the out Side of the Ditch and that by a Superiority of Cannon, and a great loss of their Troops, which last must be inevitable cost them, and that from this Place they Will be able to make a Breach, & not before (except in the Parapet) which will not be Sufficient for them to make an assault; then, and not till then, a brave officer ought to think of Capitulating, when he may reasonably expect an honorable One, for his former gallant behavior; and it's generally, if not always, that such a defense meets with great respect even from an Enemy; and they will not think it a prudent Scheme to force a brave officer to be desperate, being convinced from his former Conduct he Will make their attempt cost dear, this manner of acting must reflect honor on the Commandant & Garrison, and no doubt but it Will bring him a timely relief, or procure him Honorable Conditions.

Scouts should be always kept out to give timely Notice, and Sally's during the Siege should be as often attempted as times and Seasons will admit; but they should be made with the utmost precaution and Secrecy, otherwise they may be cut off, so Weaken the Garrison, & by that Means Shorten the Siege.

Every Materiel that can Mend Carriages, Ramparts, & Parapets, ought to be brought into the Fort, otherwise the Ramparts & parapets will Soon not be tenable, and the fire of the Cannon too soon be lessen'd: besides Spare Planks for repairing Platforms; a certain Number should be fix'd on for this Service. The honors of War are colors flying, Drums a beating, with one or two Pieces of Cannon & Match lighted & so many Rounds, and Days provisions; and the whole to march thro the Breach; But this is never allow'd to any, but those, who make an obstinate defense.

Will: Eyre, Eng^r.

In case that the Commandant is acquainted that a Body of Troops are on their March Without Cannon, He may be assured their Intentions are to approach the Garrison unobserved, and to Storm the Fort by Escalade, which is often Successful, if the People Within have not good look out, and reflects great honor on the assailants, arid the Contrary on the Garrison if, they should be Successful, but if this designed Attack be discovered by the defenders it cannot be Successful if the Commandant and his Troops do their Duty, and consequently must be fatal to the Enemy: this is one of the most Bloody attacks made against a Fortress, and fatal when the Issue is not favorable to the assailants, when this is apprehended all the Guns on the Flanks should be loaded With Grape Shot, as they being chiefly useful on Such Occasions. The Footstep all round the Ramparts should be in good repair, that every Part might be full Mann'd. If small Brush-Wood can be found a few fascines and Gabions should be made upon Notice that the Enemy are making preparations for a Siege, they being of the greatest Use to repair the Parapets, I mean the fascines fasten'd with pointed Sticks, and the Gabions, by filling them with Earth, Serve Many Purposes, but particularly in making Blinds or Traverses on any Part of the Works, which are Secur'd by the Besiegers Cannon.

One third Part if not the half of the Troops, should be on Duty at once, and to be relieved Just before Night during the Siege. The Small arms to keep a Constant firing both Night and Day, but particularly in the former, which time the Cannon should cease except the Enemy were making an attempt by Escalade; but the Mortars are to be used at all times; this Method Will render the Enemy's progress under the Shelter of Darkness very hazardous, as well as Slow which otherwise they would make use of to their advantage. A Proper Party should be posted in the advanced Work in order to keep the Enemy from making a Lodgment close to the Bank and a field Piece may be advantageously Used there, taking care that when there appears apparent Danger of its soon falling into the Enemy's Hands to be brought into the Garrison. It's impossible to enumerate all the Incidents that happen in a Siege, in order to give Directions thereupon, therefore those must be left to the Discretion and Abilities of the Commanding officer.

DIRECTIONS TO THE COMMANDANT AT FORT EDWARD

By Captain William Eyre
Albany, December 2, 1755
(*Johnson Papers*, vol. 2, 365–66)

Directions to be observed by the Commanding officers at Fort Edward in case of an attack. Three of the Six pounders should be mounted as Soon as possible in the North East Bastion two in the North west, & one in the South East, the South west Corner of the Fort to be laid out as the chief Engineer

Appendix B

has marked it and put into a Posture of defence, as Soon as time will allow, and palisaded as the rest of the Works. If an Enemy should attempt this Place, it's reasonable to believe they Will do it in those two Sides that are not defended by Water, consequently the greater care must be taken to have as many embrasures made in those Bastions and Platforms, which may enable the Garrison to fire upon the Enemy let them approach it which Way they Will, great care should be taken to oblige the Enemy to begin their approach as far off as is possible, by keeping a Constant fire on them With as Many of the Guns as can be brought to bear and particularly before they can have time to cover themselves. After they have erected their Battery or Batteries against the Garison they Will endeavor to destroy and knock down the Top of the Parapet & Palisades in order to make a Passage which they may assault, & this Step must be wholly left to the Commanding Officer, as he Will Be the best Judge of his own Strength, & that of the Enemy; but if he finds himself able to oppose the beseigers, he ought to make a Retrenchment behind the Place where he expects the attack in case he should be obliged to give Way, in order to save the Garison from being Sacked or put to the Sword, this Retrenchment is only a Breast Work raised, to retire behind if the Breach or Passage cannot be defended. But If the Enemy should endeavor to make themselves Masters of ye Fort by Escalade, it must be by Surprise, otherwise it's a very hazardous attempt, therefore the Commandant ought to be on his Guard to prevent such an attack, but if they should undertake it by mere force, the Artillery should be all loaded with Grape Shot, on the Flanks, and every Part of the Works Manned as Well as the Number Will admit except a proportional Number on the Parade to be always ready to Sustain that Part which may be pressed most. Small Parties of two or three should be constantly kept out to give timely Notice of an Enemy's approach. If the Commanding officer finds, after he has done his utmost to defend the Garison as long as he can, his next endeavor is to obtain honorable Conditions, the honours of War are that ye Garison March out With Drums beating Colours flying, two or three Days or more of Provisions, as also one or two Cannon, & Match lighted. during the time that the River is lock'd up by Ice, great caution should be used to prevent an Escalade, as an Enemy may then approach it on all Sides with ease.

WILL: EYRE.   Engineer.

Appendix B

Remarks on Forts William Henry and Edward

By Captain Harry Gordon
June 22, 1756
(Pargellis, *Military Affairs in North America*, 177–80. Note: See maps on pages 90 through 94.)

Remarks upon the Forts of William Henry and Edward of their Situation and what Works are most necessary to be added for the Strengthening of them—by Order of H.E. General Shirley.

Fort William Henry is situated at the South End of Lake George formerly called Lake St. Sacrement—It is a Work that consists of 4 Bastions with intermediate Curtains—and a Ditch eight foot deep and about thirty wide from the North-West Bastion to the South East one. The Work of the Ramparts and Parapets is faced up with large Logs of Timber bound together with smaller ones. The Rampart is in most Places fifteen Foot broad on the Curtains—the Bastions are filled up—The Parapets are, in the Faces of the Bastions most exposed, from fifteen to eighteen Foot thick, and on the Curtains from twelve to fifteen—The Rampart is between ten and eleven Foot high, and the Parapets from five to five and a half—There are Barracks for between three and four hundred Men—A Casemate under the left Flank of the South East Bastion, and another under the East Curtain. Likewise a Magazine under the N.E. Bastion towards the Lake and another smaller under the N.W. Bastion.

This Fort stands upon a high sandy Bank twenty Foot above the Lake which covers one Front—A Morass another which winds within fifty yards of the third; so that an Attack cannot be well carried against any but the Western Front. There is a rising Ground about 300 Yards distant before the South West Bastion which rises to between sixteen and eighteen Foot higher than the Ground the Fort stands upon—likewise the rising ground across the morass is higher.

In order to strengthen this Fort it is necessary to raise the Faces exposed to the rising Grounds three Foot higher—to cover and defend the South West Bastion and Curtain, from the Batteries an Enemy might raise upon the rising Ground, so as not to be battered in breach from thence—To effect this a Ravelin ought to be raised before the said Curtain, and a Counterguarde before the S.W. Bastion. A Communication ought to be made to the Ravelin—which ought to be sunk under the Curtain to come out at the bottom of the Ditch—and to cross it by a Caponiere with steps up to

Appendix B

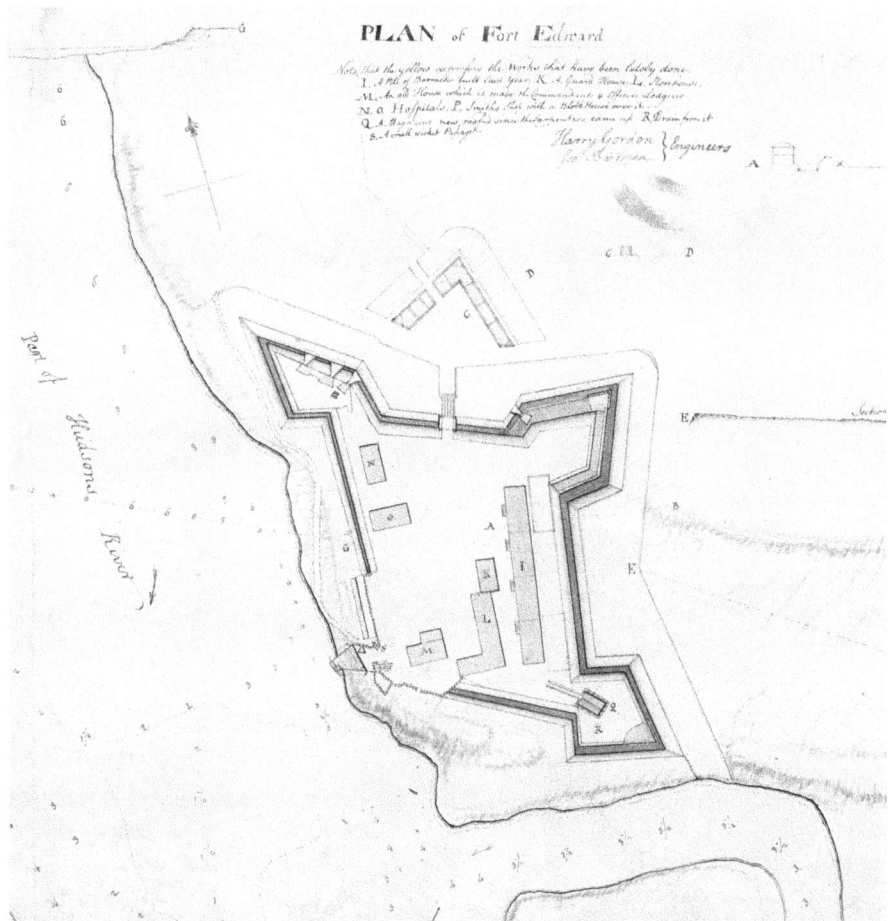

A plan of Fort Edward drawn by Captain Harry Gordon of the Royal Engineers in the spring of 1756. Even at this stage, one can see the size of the fort. The defenses of Fort Edward, considered the key to the New York frontier, would continue to be expanded and modified over the course of the war. *From Hulbert,* The Crown Collection of Photographs of American Maps, *Series II/1 (1910).*

ascend the Ravelin—A covered Way pallisadoed ought to be carried from the left Face of the Counterguard to a detached Redoubt, made last Year, very properly to scour the Bank above the Morass which was not seen by the Fort—This Redoubt for Want of the Communication being properly secured, is at present unsuitable, but may be made very necessary to scour the left Face of the Countergarde.

These proposed Works will entirely cover the exposed Front of the Fort (and without them a Breach may soon be made without shifting the Batteries

Appendix B

from the rising Ground—but if these Works are added the Enemy must first destroy them and afterwards make their Batteries in them to make a Breach in the Bastions. A Casemate should be made under the left Face of the Ravelin which cannot be battered but obliquely. The covered Way will serve for a small retrenched Camp, or a Cover for Magazines of Provisions & ca.

Fort Edward is situated on Hudsons River 14 Miles below the other Fort above described. It is a Work of four Bastions as the other—that on the River below is rather a half Bastion, one Side is close to the River another to a small Rivulet which winds towards the third. The Gate is in the Curtain towards the Plain. There is a Gate likewise in the Side that's towards the Rivulet. There is a Ditch on the North and East Sides, and a Row of Pallisades (which has been the Preservation of the Fort) goes quite round between the Ditch & the Parapet—with their Points inclining towards the Country. There is no Rampart to the Fort and the Parapet is not above eight Foot thick in some Places it has washed to six a Top. The Parapet is from eight to ten Foot high reared up of Sand, without any regular Banquet—or any kind of facing.

There is a Magazine in the East Bastion, which is only covered with one layer of Logs. The River Hudson divides itself a little above the Fort and forms a large Island opposite to it. The Branch of the River between the Fort and the Island is about sixty Yards across. The Island a hundred, and the other Branch seventy.

In order to strengthen this Fort the Parapets ought to be faced with Logs as at Fort William Henry, and made from 14 to 16 Foot thick—the Rampart on the East & South Sides ought to be raised so as to have Casemates under the Curtains—and proper Cover for 2 Magazines under the 2 Bastions—A Ravelin constructed before the Gate of the North Curtain—and a Redoubt detached before the East Curtain to discover the Banks of the Morass which are high—this Redoubt to communicate by a Sally Port under its Curtain and a covered Way well pallisaded—a covered Way may be carried from the Redoubt to the Ravelin and prolonged to the River. A Hornwork ought to be made in the Island with its Lunette across the Western Branch. This Work will secure the Passage of the River and cover Storehouses to lodge Provisions & ca. Care must be had to raise the Floors of the Storehouses as the River has been known to rise over the Island. Landing Places must be made for Boats in the Island. The Curtain towards the River must be secured against Floods as the Ground the Fort stands upon is rather lower than the Island. A small Redoubt may be made across the Rivulet the better to Flank the Hornwork.

Appendix B

These Works as the Timber is nigh may be soon Constructed, and without them the Passage of the River (The Design of this Fort) cannot be covered properly for communication nor prevented our Enemies as they may go along with any Number of Battoes or Canoes down the Western Branch without being discovered by the Fort. If it is supposed ever to be attacked the Out Works will add greatly to the Strength of it—seeing, in such Case it would have all the upper Inhabitants of the Province of New York to defend it—whose principal Frontier this Fort certainly is—and with the addition of these Works, it could with great Numbers & Risque only, be invested.

As to the Works to be added to Fort William Henry—they seem to me so necessary for a Defense—that without them the Enemy can in one Night open Trenches make a Battery within 280 Yards of the Bastion which entirely commands it and which without shifting may soon make a Breach.

Harry Gordon
Engineer

State of the Works at Fort Edward

By Colonel James Montresor
August 12, 1757
(*Montresor Journals*, 39–40)

N.W. Bastion is raised up to its height within. The Merlons filled in and Platforms laid. The Ramp up to it & its Terreplein to be made good and Levelled. NE Bastion Completed and finished. The Curtin between those two Bastions. The Rampart is not quite filled in above the Casemate, one day more will finish it.

S.E Bastion finished but wants to be Cleared & level'd The Curtin between the S.E & Water Bastion, the Parapet not quite fill'd in nor the Platforms laid. S.W Bastion or Water Bastion not raised nor its adjoining Curtin which is without Rampart & only a Fascine parapet next the River.

The Ditch of the East Curtin is finished to its proper Depth. The Ditch of the N Curtin finished. The Ditch of the S Curtin not Carried round. No Ditch to the West Curtin being next the river. The Ditch of the Ravelin will be finished tomorrow evening.

Appendix B

The Ravelin is raised to its proper height with Platforms. Its gate G, and Draw bridge wanting. The Bridge from the Ravelin to the Fort finished, the Ballances of the Draw bridge wanting. The Face of the N West Bastion wants a log the whole length to be raised under the Fraise & its Salient angle too low in front and in danger of being Brusqued.

The Casemates to be cleared & made fit for the men, there being molasses and Rum in one, Provisions in the other, Shells and ordinance Stores in the 3rd which wants to be lined next the Barracks. The 4th empty.

The officers Barracks to be finished and the men's put in order. A well and oven necessary for the Garrison. A safe Place for fixing Shells &c for the Artillery.

Memoir on Fort Carillon

By M. de Pont le Roy, engineer in chief
1758
(*N.Y. Col. Doc.* 10, 720)

This fort is built on a rock on the left bank of the River of the Falls, commanding its outlet into the River St. Frederic as well as that of the head of the bay.

It is an irregular square, the long sides of which are fifty-four toises of exterior Polygon; the small twenty-nine. Its revetement consists of squared pieces of oak laid one on the other, bound by traversines fastened to *corps morts*; its periphery is pierced with embrasures lined with oak timber and directed towards different points of the exterior ground. Only one or two guns can be opposed from the fort against all the batteries constructed by the enemy.

The ramparts are but thirteen or fourteen feet wide, and the platforms consequently so short that the recoil at each discharge makes the gun run off. Should one be dismounted, it *becomes* necessary to fire those next it, in order to convey another there.

The bastions are casemated and serve for the bakery, cistern, powder magazine and provisions. The casemate under the curtain of the entrance, which may serve to lodge the garrison, is only twelve feet wide by six high, extremely damp, the roof consisting only of beams laid side by side, covered with four or five feet of earth.

Appendix B

The place of arms is only eighteen toises long by nine wide.

The foundation is solid rock; the buildings for civilians are of stone and two stories high. The roof overtops entirely the parapets of the rampart. The shot and shell directed against these buildings; would prevent, by their explosion, the appearance of the garrison either on the place of arms or on the rampart.

The great number of embrasures excludes the use of musketry, the only means, nevertheless, of defending the place.

On the two fronts which are open to attack, a half-moon has been constructed so high that it entirely covers the embrasures of the curtains.

The covered way is not yet commenced, and part of the counterscarps remain to be built, as well as the parapets of the place on two fronts.

The cistern contains only fifteen thousand quarts of water; it is filled by the conductor from the place of arms, which has no cistern, a circumstance that renders the water muddy and no doubt unwholesome.

The powder magazine being roofed only by beams laid side by side, covered with earth, is always damp in spring and fall; the powder has to be removed.

All the store-houses and sheds, necessary for the garrison, are outside the place, encircled by a palisade.

From this description 'twill be seen how little susceptible of defence is this fort; yet, 'tis the only work that covers Lake Champlain and, consequently, the Colony. Were I entrusted with the siege of it, I should require only six mortars and two cannon.

Remarks on the Situation of Fort Carillon and Its Approaches

By Captain D'Hughs
1758
(*N.Y. Col. Doc.* 10, 707–10)

Fort Carillon stands on a rock in a tongue of land formed on the West by the waters of the Falls of Lake St. Sacrament; on the Southwest by the Bay; and from the South to the North it is bounded by the river which leads into Lake Champlain. The fort is to the Southwest of Fort Frederic. 'Tis an irregular square, the defects of which proceed from its having only between forty-five

Appendix B

and fifty toises front, instead of 80 or 100 toises at least, which it ought to have been allowed; in that case ground sufficient would have been occupied, whence several gullies whereby the enemy can approach unperceived very near the fort, might be exposed and flanked, and the English barges which would pass the river in front of the fort, might also be more readily discovered and battered at a shorter distance. This would obviate the construction of the redoubt projected to be built for that purpose on the crest of the point Southeast of the fort, which will remedy only one defect at that point and not at the others, where 'twill be necessary to construct for that purpose, besides the ordinary works, advance fortifications which will never be productive of so good an effect and will cost the government as much as the enlargement of this fort; the latter is not impossible, and would afford sufficient room to build all the King's magazine inside, some of which are outside and exposed to be burned by any one man sufficiently bold, whom the enemy may send during the drifting of the snow (*poudreries*) so frequent an occurrence in Canada, when a soldier on guard does not see twenty paces in front of him.

The site is very good and susceptible of an excellent fortification; is favorable in so far that the enemy can open the trench only on one side, where, even, he can be deprived of that advantage by removing down to the bare rock the trifle of earth that is lying on it.

The forts in this country are ordinarily constructed only of pieces of timber one over the other, in which cannon effects a practicable breach with more difficulty than in stone; therefore the forts such as are now in Canada have been, and will be taken only by force of shell; this would not be the case had not the bad habit prevailed of building forts too small at points where a place capable of resistance was required. The reason ascribed here for this is, that the troops are not numerous enough to garrison them; this reason could avail in past times only, either because the troops being in fact very few, or a train of artillery never having been seen in the field, 'twas necessary only to protect the place against a surprise.

West of the fort, at a distance of five hundred toises, is a steep hill having several faces; it borders on its left the River of the falls, and on its right forms a very steep curtain which commands a plain of 700 toises that terminates at the Fort Frederic river. The enemy inclined to besiege Carillon must necessarily render himself master of that eminence in order to cover the landing of his artillery in a cove at its foot, being unable to have it brought by land or by another side of the river, as this is the only place along the shore not exposed to the fort. It is this eminence which 'tis essential to secure, and a General desirous of preventing the siege,

Appendix B

must have a good intrenchment erected on it, which he must even have continued across the plain as far as the Fort Frederic river. That line, 1000 toises in length, forms the base of the angle on which Carillon stands. This intrenchment of trunks of trees to be felled at the moment they are required, must be fraised with dry branches well lopped and entangled together; the approaches to it ought to be encumbered by that abatis for a distance of fifty toises, observing particularly that no large trunks of trees be piled up at the extremity.

Whatever need there be of wood on other occasions, it must be taken from some other quarter, and that side must be left unstripped of the trees which will be found very handy in urgent necessity. This intrenchment, which can be completed in twice twenty-four hours, and well-guarded by six thousand men, would cost the party desirous of forcing it, a great many lives, and I even dare assert that, were it well defended, 'twould not be carried by an army three times more numerous than that defending it. This work is already begun on the Northeast of the fort, by an abatis of about 400 toises which was constructed three years ago, when digging a trench (*trancé*) down to the river of Fort Frederic in order to be able to destroy the bateaux and sloops coming from that direction.

Half a league West of the fort is a considerable Fall of the waters of Lake St. Sacrament. Those going into that lake commence at this point, a portage of half a league to another little fall at the mouth of said lake. This last fall is called "The Portage," and is a very favorable post for an army strong enough to oppose, at all points of disembarcation, the landing of an enemy's force, but also to detach a strong body of troops in order to oppose that portion of this hostile army which may come by land to cover that descent, and cut off all our retreat and communication with Carillon; to effect this it need only occupy the post of the Great Falls; in the case I cite, the post of the portage is, I repeat, a good one, because it is the only place, in approaching by Lake St. Sacrament, where the enemy can land their artillery to convey it, afterwards, to the Great Falls and to bring it thence by water before the fort near that hill I have already mentioned.

Convinced of the absolute necessity that exists in this Colony of solely and permanently securing this frontier, 'tis astonishing that Carillon has not been made a large and strong place, susceptible of a long resistance. I would be still of the opinion to construct, at the mouth of Lake St. Sacrament, in the vicinity of this portage, at the only place where artillery can be landed, a strong redoubt or little fort capable of resisting every attempt to escalade it, on the part of the enemy who, being unable, until after the capture of that little fort, to land any artillery, could not make use of it except at a great distance, and on pontoons which could be sunk, and from which the guns could not be well aimed.

South of Carillon is a large bay extending about nine leagues inland, which conjointly with the waters of the Fall forms, in front of this fort, the river of Fort Frederic. 'Tis by this bay the enemy come often to scout in barges, and some of them are constantly passing, under cover of the night, into Lake Champlain, where they come to intercept our convoys and capture voyageurs' bateaux when passing few in number; six leagues from Carillon, in this bay, is a narrow pass called "The Two Rocks," which furnishes a very advantageous position for another little fort similar to that which might be constructed at the portage.

By means of these two little forts the enemy would be prevented disquieting us in any way in our communication with Montreal, and whenever disposed to come and lay siege to Carillon, would be stopped sufficiently early, in front of these outposts, to afford us time to throw into the principal place all the succors possible and to seize our advantages to enable us to fight his army more safely. To march a train of Artillery in Canada is a matter of considerable trouble and difficulty; therefore this description of redoubts in advance of, and not far from a strong-hold, especially when located at the mouth of a lake and at the only place for landing, would stop, for a long time, an army of this country. By these obstructions, a General might lose the best time of his campaign, which cannot be long in this climate, and would not be able, in one summer, to reach the walls of Montreal as he can do, having Carillon only to take, especially should he arrive there before we had time to oppose him with an army which, in this country, cannot assemble as diligently as in Europe, where there are no contrary winds to be dreaded in ascending lakes and rivers.

Fort Carillon once taken, Fort Frederic would not stand an instant. The latter is built of stone so as to be incapable of resisting four cannot shot, which would be sufficient to tumble it utterly into ruins. All the country in its vicinity is flat and affords, at every step, an easy landing for the largest guns; even firing a few shot at it from pontoons would be enough to render it incapable of answering, it being all shook. An army of observation could not intrench itself under that fort except with earth, a work affording much poorer defence than the intrenchment I have mentioned, and requiring extremely long and fatiguing labor.

Neither would the enemy meet any impediment to his progress; neither portage, nor fort; for I do not regard as such that of Saint John, which

Appendix B

consists of upright pickets; so that he would find himself in the center of the Colony and master of this entire frontier.

The two little out-posts I propose would be of trifling expense, being fortified naturally by their positions, and could not serve against us for the reason that the enemy cannot retain them as they are at too great a distance from his principal posts and too near ours, whence we could march to retake them before the enemy had time to be aware of the fact and to reinforce them. This could be effected even during the winter.

The guarding those posts ought to be confided to the best troops of the Marine who are little inclined to desertion and regard this country as their home; the Commandant ought also pay attention to have the ice, whenever it would begin to take, frequently broken up in the neighborhood of his post to a certain distance. Hostile parties would not risk themselves too much in leaving in their rear, and so near them, these posts whence they could be cut off, whenever they would come to examine the movements making at Carillon, either for the campaign or for winter detachments, attention being constantly paid to the keeping always at these out-posts some Indians who would go out at the first signal made at Carillon, the moment information would be received of any party of the enemy.

D'HUGUES.
Carillon, 1st May, 1758.

Appendix C

THE LEGEND OF DUNCAN CAMPBELL

No telling of the 1758 Ticonderoga campaign would be complete without the legend of Duncan Campbell. One of the oldest ghost stories in North America, the following, as related by Campbell family tradition and told by Highland historian Dean Stanley, is taken from Francis Parkman's Montcalm and Wolfe *(Cambridge, MA: Da Capo Press, 2001, 433–35.)*

The ancient castle of Inverawe stands by the banks of the Awe, in the midst of the wild and picturesque scenery of the Western Highlands. Late one evening, before the middle of the last century, as the laird, Duncan Campbell, sat alone in the old hall, there was a loud knocking at the gate; and, opening it, he saw a stranger, with torn clothing and kilt besmeared with blood, who in a breathless voice begged for asylum. He went on to say that he had killed a man in a fray, and that the pursuers were at his heels. Campbell promised to shelter him. "Swear on your dirk!" said the stranger; and Campbell swore. He then led him to a secret recess in the depths of the castle. Scarcely was he hidden when again there was a loud knocking at the gate, and two armed men appeared. "Your cousin Donald has been murdered, and we are looking for the murderer!" Campbell, remembering his oath, professed to have no knowledge of the fugitive; and the men went on their way. The laird, in great agitation, lay down to rest in a large dark room where at length he fell asleep. Waking suddenly in bewilderment and terror, he saw the ghost of the murdered Donald standing by his bedside, and heard a hollow voice pronounce the words: "Inverawe! Inverawe! Blood has been shed. Shield not the murderer." In the morning Campbell went to the hiding place of the guilty man and told him that he could harbor him no longer. "You have sworn on your dirk" he replied and the laird of Inverawe, greatly perplexed and troubled, made a compromise between conflicting duties, promised not to betray his guest, led him to the neighboring mountain and hid him in a cave.

In the next night, as he lay tossing in feverish slumbers, the same stern voice awoke him, the ghost of his cousin Donald stood again at his bedside, and again he heard the same appalling words: "Inverawe! Inverawe! Blood has been shed. Shield not the murderer!" At break of day he hastened, in strange agitation, to the cave; but it was empty, the stranger had gone. At night, as he strove in vain to sleep, the vision appeared once more, ghastly pale, but less stern of aspect than before. "Farewell, Inverawe!" it said; "Farewell, till we meet at TICONDEROGA!"

The strange name dwelt in Campbell's memory. He had joined the Black Watch, or Forty-Second regiment, then employed in keeping order in the turbulent Highlands. In time he became its major; and, a year or two after the war broke out, he went with it to America. Here, to his horror, he learned that it was ordered to the attack of Ticonderoga. His story was well known among his brother officers. They combined among themselves to disarm his fears; and when they reached the fatal spot they told him on the eve of the battle, "This is not Ticonderoga; we are not there yet; this is Fort George." But in the morning he came to them with haggard looks. "I have seen him! You have deceived me! He came to my tent last night! This is Ticonderoga! I shall die today!" and his prediction was fulfilled.

Appendix D

THE FORTS TODAY

It is surprising how many of the forts discussed in this text still stand today in one form or another. This is an impressive feat given the seasonal ravages of over two and a half centuries. The ruins of both Fort Chambly and Fort Ticonderoga were reconstructed, and today, both appear as they would have in colonial times. A reconstruction of Fort William Henry is to be found at the head of Lake George, which captures the same feel and window into the past that is found at the previous mentioned locations. If the reader should have the opportunity to visit any of these locations, I would wholeheartedly encourage it.

While these forts stand intact, the remaining colonial structures are now ruins. The remains of Fort St. Frederic and Fort Amherst are at Crown Point State Park. Spend a few minutes at this scenic location and it is easy to understand why it took on such a strategic importance. Fort St. Jean would be rebuilt and destroyed again in the American Revolution, and Île aux Noix would see a host of fortifications raised during the American Revolution, the War of 1812 and beyond. Today, Fort Lennox National Park occupies the site.

Fort Edward and the forts at Albany are gone as well, as are all the small posts that tracked the northern shores of the Hudson River. Even the series of forts built at the tip of Manhattan have vanished, eventually being replaced by the Alexander Hamilton U.S. Customs House, which stands at the location today. Rumor has it, however, that not all involved agree with the change. Parishioners at nearby St. Mark's Cathedral, where Governor Stuyvesant was interred in 1672, have been known to hear a man with a wooden leg walking about, perhaps looking for the ramparts of his fallen fort.

NOTES

Introduction

1. Hadden, *Journal*, 85, 88–89; "A Journal of Carlton's and Burgoyne's Campaigns," *FTMB* 11, no. 6 (September 1965): 312, 321.
2. Pouchot, *Memoirs of the Late War*, 78.
3. This point, and the need to cut the enemy's communications, led to a common recommendation that the besieger outnumber the defenders by at least three to one.
4. Pargellis, *Lord Loudoun in North America*, 94–95.
5. Kimball, *Correspondence of William Pitt*, I, 145, 194, 231, 252–55; Godfrey, *Pursuit of Profit and Preferment*, 116–18; "Diary of Benjamin Glasier," EIHC, 86 (1950): 71, 75.
6. Pargellis, *Lord Loudoun in North America*, 298–99; "Mismanagement: The 1758 British Expedition Against Carillon," *FTMB* 4 (1992): 254; Perry, *Recollections of an Old Soldier*, 9.
7. *NY Col. Doc* 10, 604–5, 621–26; Bougainville, *Adventure in the Wilderness*, 175–77.
8. Laramie, "French Lake Champlain Fleet," 1–32.
9. Haviland's Journal, 3–4 Sept., 1760, WO34/77; Jenks, "Journal," 375–76; List of Prisoners taken...at Ile-aux-Noix and Chambly, 18 Oct., 1760, CO5/59.
10. *JP* I, 483–84.

11. For a detailed history of Fort Montgomery, see James Millard's *Bastions on the Border: The Great Stone Forts at Rouses Point on Lake Champlain* (Burlington, VT: America's Historical Lakes, 2009).

Chapter 1

12. *Jesuit Relations* 22, 247, 275–81; *Le Bulletin des Recherches Historiques* (January 1948): 8–9.
13. Jacques Rabelin de la Crapaudière.
14. *Le Bulletin des Recherches Historiques* (January 1948): 8–10; *Jesuit Relations* 30, 183. An interesting account of the life and career of Jean Bourdon can be found in Reverend M.W. Burke-Gaffney's paper titled "Canada's First Engineer Jean Bourdon (1601–1668)," *Canadian Catholic History Association* 24 (1957): 87–104.
15. *Jesuit Relations* 49, 161–63, 237–39, 253; Verney, *Good Regiment*, 19, 21–22.
16. Verney, *Good Regiment*, 29; *Jesuit Relations* 49, 266; *Guide to Fort Chambly*, 6; Beaudet and Cloutier, *Archaeology at Fort Chambly*, 35. Archaeological evidence has called into question the historical image of Fort St. Louis. Although only partial traces of the fort's walls have been uncovered, there is no sign of Captain Chambly's redans. Instead, the original structure seems to have been constructed along the lines of a bastioned structure similar to Fort Richelieu. (*Archaeology at Fort Chambly*, 33–35.)
17. *Jesuit Relations* 49, 165, map facing page 266; *Jesuit Relations* 50, 81–83; Verney, *Good Regiment*, 21, 26.
18. Memoire de M. de Salières des choses qui se sont passees en Canada…1665–1666, MG7-IA2, vol. 4569, fol. 98–102.
19. Verney, *Good Regiment*, 27–31; *Le Bulletin des Recherches Historiques* (January 1948): 13–14; *Jesuit Relations* 49, 255, map facing page 266.
20. Vaudreuil to Minstre, November 12, 1712, MG1-C11A, vol. 33, fols. 15–37.
21. Memoire anonyme, November 6, 1702, MG1-C11A, vol. 20, 253–54; Beaudet and Cloutier, *Archaeology at Fort Chambly*, 11–12, 41–42, 51–69; *NY Col. Doc* 9, 841, 846, 851; Vaudreuil to Minstre, November 12, 1712, MG1-C11A, vol. 33, fols. 15–37; Memoire de M. de Lery—fort de Chambly, October 26, 1720, MG8-A1 series 3, vol. 7, 741–44. "The value of the fort," de Lery wrote in an assessment, "resides above all in the strength of its garrison, but this had always been one of its major deficiencies." (Beaudet and Cloutier, *Archaeology at Fort Chambly*, 12.)

22. Beauharnois et Hocquart au Ministre, October 5, 1740, MG1-C11A, vol. 73, 19–20v; Projet de dépense pour la construction et armement d'une gabare ou bateau plat…à faire naviguer dans le lac Champlain, ibid., vol. 73, fol. 21–22; Lery to Ministre, October 26 and November 7, 1744, ibid., vol. 82, fol. 296–303v, fol. 306–7v.
23. *NY Col. Doc.* 10, 180–81; Lery to Ministre, October 28, 1748, MG1-C11A, vol. 92, 284–89v; Kalm, *Travels into North America*, III, 50; Pouchot, *Memoirs on the Late War*, 337. Louis Antoine Bougainville, a French officer who had to travel the St. Jean to La Prairie road a decade later, referred to it as abominable. "Until they give it a place to drain off," he noted in his journal, "the road will always be impractical." (Bougainville, *Adventure in the Wilderness*, 65.)
24. Galissonniere et Bigot to Ministre, September 26, 1748, MG1-C11A, vol. 91, fol. 40–45v; Kalm, *Travels into North America*, III, 45–46.
25. Franquet, "Voyages et mémoires sur le Canada par Franquet," Institut Canadien de Québec, *Annuaire* 13 (1889): 61–63.
26. Casgrain, *Lettres de Bourlamaque*, 10–17, 26–28; Levis to Bourlamaque, May 23 and June 2, 1759, NAC MG18-K9, vol. 3, 83–86, 91–94; Charbonneau, *Fortifications of Île aux Noix*, 20–24.
27. Laramie, "French Lake Champlain Fleet," 1–32.
28. Webster, *Journals of Jeffrey Amherst*, 181–83.
29. Casgrain, *Lettres de Divers Particuliers*, 137–47; *NY Col. Doc* 10, 1101–02; Charbonneau, *Fortifications of Île aux Noix*, 24–35, 41–49; Johnstone, *Memoirs of the Chevalier de Johnstone*, III, 68–69.
30. Casgrain, *Lettres de Divers Particuliers*, 148–49; Mémoire sur la Frontière du Lac Champlain par M le Chevalier de Bourlamaque, NAC MG18-K9, vol. 6, 105–15; Examination of Prisoners aboard the *Duke of Cumberland*, June 20, 1760, WO34/51; *Pennsylvania Gazette*, September 11, 1760; Charbonneau, *Fortifications of Île aux Noix*, 331–36; *NY Col. Doc* 10, 1104.
31. Rogers, *Journals*, 192–94; Haviland's Journal, August 30–31, 1760, WO34/77; "Journal of Sergeant David Holden," 399–400, Haviland's Journal, September 3–4, 1760, WO34/77; Jenks, "Journal," 375–76; List of Prisoners taken at Île-aux-Noix and Chambly, October 18, 1760, CO5/59. The fort was hardly in a position to resist, but it seems that in keeping with custom, Lusignan believed that by invoking the English to fire he could ask for more honorable terms. Darby refused these and informed Lusignan that if he did not surrender unconditionally he would put the garrison of seventy soldiers to the sword. Jenks recorded in his journal an unusual and frightening event during the brief siege:

"Our people took some of the inhabitants—women and children—and placed them before their royal (5½" mortar), and so fired over their heads, which answered instead of fascine batteries." (Jenks, "Journal," 375.) It is questionable whether a senior British officer like Darby would have allowed such an action to occur; news of this would have greatly damaged his reputation, not to mention the fact that it ran contrary to Haviland's orders that the general populace was not to be molested or harmed on pain of death.

Chapter 2

32. *NY Col. Doc.* 9, 400, 1021–23; Roy, *Hommes et Choses du Fort Saint-Frederic*, 9–13. At this time, Pointe-a-la-Chevelure or Scalp Point referred to the peninsulas on both sides of the lake. Today, these are known separately as Crown Point and Chimney Point. The English and Dutch also did not distinguish between the two sides of the lake at this time, referring to the whole region as Crown Point, with the term "crown" referring to the top of one's head.
33. *NY Col. Doc.* 9, 1021–25.
34. Roy, *Hommes et Choses du Fort Saint-Frederic*, 20–21; Plan du Terrain de la Pointe a la Chevelure, October 25, 1731, MG1-C11A, vol. 54, fols. 346–46v; Conseil de Marine à M. Rocbert de Morandière, April 22, 1732, MG1-B, vol. 57, fol. 644; *NY Col. Doc.* 9, 1034, 1037. For the initial colonial reaction to the French fort, see *Cal. A&WI* 38 (1731): 312–14, 331–32.
35. Beauharnois et Hocquart au Ministre, November 14, 1731, October 1 and October 14, 1733, MG1-C11A, vol. 54, 338–42, 385–86v, 423–24v; Memoire de Chaussegros de Lery, October 25, 1731, ibid., 344–45; Memoire du Roi…, April 22, 1732, MG1-B, vol. 57, fol. 652. Also Roy, *Hommes et Choses du Fort Saint-Frederic*, 22–32, which contains portions of the above correspondence and excerpts of several additional letters.
36. Lery au Ministre, October 30, 1735, MG1-C11A, vol. 64, fols. 259–61.
37. Work continued on the structure for several more years. The covered way was improved, as was the garrison's accommodations and the outer ditch. In 1742, a dockyard was added to accommodate a barque and additional works placed to defend it. (Beauharnois et Hocquart au Ministre, October 5, 1740, and October 31, 1742 (two letters), MG1-C11A, vol. 73, 19–20v, vol. 77, 72–73v, 141–42; Etat estimatif des ouvrages necessaires a faire

au fort Saint-Frederic, October 10, 1740, ibid., vol. 73, 24–25v; Lery to Ministre, October 30, 1742, ibid., vol. 64, fols. 259–61.
38. Franquet, "Voyages et Memoires," 165.
39. Ibid.
40. Kalm, *Travels*, III, 36–37; Franquet, "Voyages et Memoires," 164–66.
41. Michel Chartier, sieur de Lotbiniere, "Vaudreuil to Lotbiniere, Sept 20, 1755," *FTMB* (January 1928).
42. Ibid.; Lotbiniere to Minister, October 31, 1756, MG1-C11A, vol. 101, fols. 333–34. Mount Independence was referred to as Pointe au Diamant at the time, while Rattlesnake Mountain was a literal translation from the French Serpent a Sonnette.
43. Lotbiniere to Minister, October 31, 1756, MG1-C11A, vol. 101, fols. 333–34.
44. "Michel Chartier de Lotbiniere the Engineer of Carillon," NYHM (1934) 32; Lotbiniere to Minister, October 31, 1756, MG1-C11A, vol. 101, fols. 333–34; *NY Col. Doc.* 10, 493–494; Rogers, *Journals*, 8–9.
45. Lotbiniere to Minister, October 31, 1756, MG1-C11A, vol. 101, fols. 333–34; "De Lery Journal," *FTMB* (July 1942): 128.
46. "De Lery Journal," *FTMB* (July 1942): 138.
47. Lotbiniere to Minister, October 31, 1756, MG1-C11A, vol. 101, fols. 333–34; "De Lery Journal," *FTMB* (July 1942): 128–44; "The Building of Fort Carillon, 1755–1758," *FTMB* (1955): 29–37.
48. "De Lery Journal," *FTMB* (July 1942): 137–38, 141–43; "Michel Chartier de Lotbiniere the Engineer of Carillon," 33–34.
49. This is, of course, a point of speculation; however, Montcalm and several of his officers were quick to recognize the potential of this position, and as such, it is difficult to picture that Lotbiniere, a military engineer who spent years at Carillon, would not have seen the same advantages. In a later letter, which Montcalm and his supporters were quick to discredit, the colonial engineer implied that he had envisioned a defensive abatis along these heights the moment the fort was threatened by a superior force. (*NY Col. Doc.* 10, 893.)
50. Lotbiniere to Minister, October 31, 1756, MG1-C11A, vol. 101, fols. 333–34; Casgrain, *Lettres du Chevalier de Levis*, 53, 68; *NY Col. Doc.* 10, 493–94.
51. *NY Col. Doc.* 10, 740, 796; *Lee Papers*, I, 12.
52. Hervey, *Journals*, 50; "James Abercrombie to Harry Erskine, 10 July, 1758," 69; General Abercromby to James Abercromby, August 19, 1758, Chatham Fonds MG23-A2, vol. 6; "Spicer Journal," 395, 407–8.

53. "Experiences in Early Wars in America," *Journal of American History* (1907): 92; Samuel Fisher's Diary, LOC; "Spicer Journal," 395, 408; Loring to ---, August 19, 1758, Chatham Fonds MG23-A2, vol. 8; *Gentleman's Magazine*, September 1758; "Journal of Lemuel Lyon," *Journals of Two Private Soldiers*, 23; *Lee Papers*, I, 14; Hervey, *Journals*, 50. Both Captain Charles Lee and Captain William Hervey claimed that Abercromby was among the first to set sail.
54. Gabriel, *Desandrouins*, 296–97; Knox, *Historical Journal*, I, 506–8.
55. Webster, *Journals of Jeffery Amherst*, 146; Colonel William Amherst's Journal, CO5/56; "Journal of the Reduction of the French Fort of Ticonderoga," *New American Magazine*, August 1759; "Captain-Lieutenant Skinner's Journal," *The Seven Years War in Canada*, 62–64; Rogers, *Journals*, 141–42; "Eli Forbush Letter," 51; Gabriel, *Desandrouins*, 297.
56. Gabriel, *Desandrouins*, 299; *NY Col. Doc.* 10, 1055. The windmill/lookout post a few hundred yards farther up the lake was also blown up.

Chapter 3

57. Today, Castle Island is known as West Island.
58. Jameson, *Narratives of the New Netherlands*, 47–48; O'Callaghan, *History of the New Netherlands*, I, 76–78. The fort's first commander was Jacob Jacobz Elkens, who remained in this post for four years. (Ibid., I, 76.) In the spring of 1639, David de Vries, spending time at a farm on Castle Island, left an account of the flooding issues that forced the abandonment of Fort Nassau a generation before: "There came such a flood upon the island on which Brand Pijlen dwelt (my host for the time being) that we had to abandon the island, and to use boats in going to the house, for the water stood about four feet deep on the island, whereas the latter lies seven or eight feet above ordinary water. This high water lasted three days before we could use the houses again. The water came into the fort. We had to resort to the woods, where we set up tents and kept great fires going." (Jameson, *Narratives of the New Netherlands*, 207.)
59. In fortification and siege craft terms, a curtain wall is a wall that connects two bastions, which typically constitutes the outer wall of a fort. For additional fortification terms, see Appendix A.
60. Pargellis, *Military Affairs in North America*, 266.
61. Father Isaac Jogues, "Novum Belgium," in Jameson, *Narratives of the New Netherlands*, 261.

62. O'Callaghan, *History of the New Netherlands*, I, 86–87, 100; Jameson, *Narratives of the New Netherlands*, 261; Steele, *Warpaths*, 112.

63. *NY Col. Doc.* 3, 255, 260.

64. "Wadsworth's Journal," *Massachusetts Historical Society Collections*, ser. 4, no. 1 (1852): 105–6.

65. *NY Col. Doc.* 3, 391.

66. *Cal. A&WI* 16, 223; *NY Col. Doc.* 4, 328–30. Romer was perhaps understating the state of the fortifications. Bellomont, who personally viewed the works at Albany and Schenectady in June, wrote the Board of Trade that "the Forts of Albany and Schenectady are so weak and ridiculous, that they look like pounds to impound cattle in, than Forts." (*NY Col Doc.* 4, 608.)

67. *NY Col. Doc.* 4, 440–41. Romer estimated the cost of the proposed works along the New York frontier as "at Albany a Fort, £4,000, at Schenectady a Fort, £4,000, at Rudgio a Fort, £6,000, at Sheractoge (Saratoga) a redoubt, £1,000, at Canestigogione a redoubt, £1,000 at the Half-moon a redoubt £1,000." The actual location of Rudgio is unclear, and the only clue come from Bellomont, who states that "Rudgio is suppos'd our northernmost between the Province of New York and Canada, as Ste. Croix is our most Eastern boundary next to N. Scotia, which is the reason Col. Romer thought those two forts would require more strength and cost than the rest." (*Cal. A&WI* 18, 691).

68. *Cal. A&WI* 16, 39–50, 55–66, 100–4; *Cal. A&WI* 32, 202; *Boston Newsletter*, September 3, 1711; Knight, *Journals of Mdm. Knight and Rev. Mr. Buckingham*. For the details behind the formation of the 1711 expedition, see Morgan and Morgan, *Queen Anne's Canadian Expedition*.

69. Brandow, *Story of Old Saratoga*, 28–29; *NY Col. Doc.* 6, 374–75; Pearson, *History of the Schenectady Patent*, 310–17; *Journal of the Legislative Council of New York, Vol. I, 1691–1743*, 391–92, 471, 631, 641. At the same time Fort Crosby was constructed, the old stone church in the center of the town was modified to act as a defensive strongpoint.

70. "Diary of Rev. Samuel Chandler," *NEHGR* 17 (1863): 346–54. The number of guns in the fort was something of an illusion. Governor Hardy reported to the Board of Trade in early 1756 that "it was with some difficulty I could furnish Fort Edward with the few now there from the Fort at Albany, and those left in it are not safe in Firing." (*NY Col. Doc.* 7, 2–3.)

71. *NY Col. Doc.* 4, 967–71, 1057, 1128–29; V, 631, 923–24, 927; *Cal. A&WI* 41, 136–42, 449–54. Captain William Hervey of the Forty-Fourth

Regiment noted in his journal that Fort Frederick mounted sixteen six-pound guns in 1755. (Hervey, *Journals*, 3.)
72. Pargellis, *Military Affairs in North America*, 259.
73. *JP* II, 99, 114, 117–18, 130–31.
74. "Journal of Captain Nathaniel Dwight," *NYGBR* (January and April 1902): 65–66; *JP* II, 328–31; Pargellis, *Military Affairs in North America*, 178.
75. "Diary of James Hill," 611–16; Pomeroy, *Journals*, 121; Dwight, "Journal," 8, 65–66 (also see Dwight's line drawing of the fort in this journal); Pargellis, *Military Affairs in North America*, 178.
76. "Diary of James Hill," 616–17; "Journal of Captain Nathaniel Dwight," 67; *JP*, II, 301, 312–13, 328–37, "Extracts from the Diary of Rev. Samuel Chandler," *NEHGR* (October 1863): 352.
77. Gabriel, *Desandrouins*, 87-88; *Bougainville Journals*, 160-161.
78. "The Journal of Adam Williamson," Williamson Family Papers, NAC A-573.
79. *Montresor Journals*, 36, 39–40; Hervey, *Journals*, 43.

Chapter 4

80. *NY Col. Doc.* 1, 190, 303.
81. Ibid., 152–53, 499, II, 16.
82. Ibid., I, 541–49; Birch, *Collection of the State Papers of John Thurloe*, I, 721–22, II, 418–19, 425. The controversial document titled "The Second Part of the Amboyna Tragedy; or a True Account of a Bloody, Treacherous, and Cruel Plot of the Dutch in America" can be found in Appendix G of O'Callaghan, *History of the New Netherlands*, II.
83. *NY Col. Doc.* 3, 51–63, 65–66; XIII, 368, 393; Fernow, *Records of New Amsterdam*, V, 88–89; Jameson, *Narratives of the New Netherlands*, 451, 461–62; O'Callaghan, *History of the New Netherlands*, II, 516–18, 520–21. Nicolls's fleet consisted of the *Guinea* (thirty-six guns), *Elias* (thirty-six guns), *Martin* (sixteen guns) and the *William and Nicholas* (ten guns). From the provisions loaded onto the vessels—five hundred matchlocks, five hundred firelocks, five hundred pistols, fifty carbines, five hundred swords, two mortars, two brass sakers and one thousand bandoliers—it's clear that Nicolls expected to raise a sizable number of New England volunteers to augment his forces. (*Cal. A&WI*, V, 189–90.)
84. Fernow, *Records of New Amsterdam*, V, 105–6; *NY Col. Doc.* 2, 372–73, 376; 13, 392–93.

85. Brodhead, *History of New York*, vol. 2, 26–31; *Cal. A&WI*, V, 225–28.
86. Brodhead, *History of New York*, vol. 2, 32.
87. Jameson, *Narratives of the New Netherlands*, 414–15, 451–52, 460–65; *NY Col. Doc.* 2, 248–53, 469. In Holland, Stuyvesant was unfairly criticized for giving up New Amsterdam without a fight. Given the state of the city's defenses, the militia's refusal to fight and the strength of the English, the governor rightly concluded that any resistance would have been senseless.
88. Paltsits, *Minutes of the Executive Council of New York* 2, 699–704, 737–39, 742–47; *NY Col. Doc.* 3, 200–1, 204–6.
89. *NY Col. Doc.* 2, 527–28, III, 199–206; *Cal. A&WI* 7, 509–11, 523–24. For his part in the affair, Manning was court-martialed and had his sword broken over his head, only having avoided a death sentence by the act of having spoken with the king upon his return to England. Given the circumstances, it is difficult to imagine what else Manning could have done to prevent the colony's collapse. (Brodhead, *History of New York*, vol. 2, 276.) The Dutch would change the name of the fort to Fort Willem Hendrick.
90. *NY Col. Doc.* 3, 589.
91. Ibid., 3, 589–91, 614–16; *Cal. A&WI* 13, 193–94.
92. *NY Col. Doc.* 4, 836–37, 929; 6, 184–86; 7, 341; *Cal. A&WI* 44, 129, 209.
93. Mention should also be made of a few other defensive positions south of Albany. Although once past New York City there were no major fortifications strong enough to bar access to the Hudson, there were a few strong points capable of repelling raiders. These locations, such as Esopus and Kingston, relied not so much on the palisade fortifications that encircled the town but on the number of militia they could muster to defend the structure. Sketches of these locations in 1695 can be found in Miller, *Description of the City and Province of New York*.

Bibliography

The following abbreviations have been used throughout this text for frequently cited sources.

NY Col. Doc.	O'Callaghan, *Documents Relative to the Colonial History of the State of New York.*
Cal. A&WI	Sainsbury, Fortescue, et al., eds., *Calendar of State Papers, Colonial Series, American and West Indies.*
DHSNY	O'Callaghan, *Documentary History of the State of New York.*
JP	Sullivan, *Papers of Sir William Johnson.*
NAC	National Archives of Canada, Ottawa.
CO5	Colonial Office Records, American and West Indies, London.
WO	War Office Records, London.
EIHC	Essex Institute Historical Collections.
NEHGR	*New England Historical and Genealogical Register.*
FTMB	*Fort Ticonderoga Museum Bulletin.*
NYHSC	New-York Historical Society Collections.

BIBLIOGRAPHY

MANUSCRIPT SOURCES

Canada. National Archives (Ottawa).

France, Archives Nationales (Paris).
6Archives des Colonies (copies in Canadian National Archives)
Sèrie C11A, Canada et Dépendances (Lettres des Gouverneurs, Intendants, officers et autres)

Great Britain, Public Record Office (London).
Colonial Office
    CO5, America and West Indies, Correspondence, originals on microfilm
War Office
    WO34, Amherst Papers, originals on microfilm

Library of Congress (Washington, D.C.).
Fisher, Samuel. Diary, 1758. Manuscript Division.

Manuscript Division
MG1: Fonds des Colonies
    C11A—Correspondance générale
MG7: Bibliothèque nationale
    IA2—Département des manuscrits, Fonds français
MG18: Pre-Conquest Papers
    K9—Bourlamaque Papers
    N21—Williamson Papers
MG23: Late Eighteenth-Century Papers
    A2—Chatham Papers

BOOKS AND PERIODICALS

Albertson, Garrett. "Montcalm's Victory." *The Bulletin of the Fort Ticonderoga Museum* IV, no. 2 (1936): 43–47.
Almon, John, ed. *A Collection of All the Treaties of Peace, Alliance, and Commerce between Great Britain and Other Powers: From the Revolution in 1688 to the Present Time.* 2 vols. London: J. Almon, 1772.

BIBLIOGRAPHY

Amherst, Jeffery. *Jeffery Amherst Papers.* Great Britain, Public Records Office, WO34.

Baldwin, Jeduthan. "Extracts from the Diary of a Revolutionary Patriot." *Journal of the Military Service Institution* 39 (1906): 121–30.

Bartman, George. "The Siege of Fort William Henry: Letters of George Bartman." *Huntington Library Quarterly* 12 (1949): 415–25.

Bayard, Nicholas, and Benjamin Fletcher. *A Journal of the Late Actions of the French at Canada...* London: Richard Baldwin, 1693.

Beaudet, Pierre, and Céline Cloutier. *Archaeology at Fort Chambly.* Ottawa: Canadian Parks Service, 1989.

Birch, Thomas, ed. *A Collection of the State Papers of John Thurloe.* 7 vols. London: Gyles, Woodward & Davies, 1742.

Blodget, Samuel. *The Battle Near Lake George in 1755: A Prospective Plan with an Explanation Thereof.* London, 1756. Repr., London: Henry Stevens, Son & Stiles, 1911.

Boston News-Letter.

Bougainville, Louis Antoine. *Adventure in the Wilderness: The American Journals of Louis Antoine de Bougainville, 1756–1760.* Trans. and ed. by Edward P. Hamilton. Norman: University of Oklahoma Press, 1964.

Brandow, John Henry. *The Story of Old Saratoga and History of Schuylerville.* Saratoga Springs, NY: Robson & Adee, 1906.

Brehm, Lieutenant Diedrich. "A New Description of Fort Ticonderoga." *The Bulletin of the Fort Ticonderoga Museum* 11, no. 1 (1962): 35–48.

Brodhead, John Romeyn. *History of the State of New York.* 2 vols. New York: Harper & Brothers, 1853, 1871.

"The Building of the Fort." *Fort Ticonderoga Museum Bulletin* 2, no. 3 (1931): 88–97.

Burk-Gaffney, Reverend M.W. "Canada's First Engineer, Jean Bourdon (1601–1668)." *Canadian Catholic History Association* 24 (1957): 87–104.

Cardwell, John M. "Mismanagement: The 1758 Expedition Against Carillon." *The Bulletin of the Fort Ticonderoga Museum* 15, no. 4 (1992): 236–91.

Casgrain, H.R., ed. *Collection des Manuscrits du Maréchal de Lévis.* 12 vols. Montreal and Quebec: C.O. Beauchemin & Fils and Demers & Frère, 1889–95. The individual volumes used in this collection are:
Vol. 1: Journal du Chevalier de Levis.
Vol. 2: Lettres du Chevalier de Levis.
Vol. 5: Lettres du M. de Bourlamaque.
Vol. 6: Lettres du Marquis de Montcalm.
Vol. 7: Journal du Marquis de Montcalm.

Vol. 8: Lettres du Marquis de Vaudreuil.
Vol. 10: Lettres de Divers Particuliers.
Vol. 11: Relations et Journaux de différentes expeditions faites durant les années 1755, 1756, 1757, 1758, 1759, 1760.
Chandler, Samuel. "Extracts from the Diary of Rev. Samuel Chandler." *New England Historical and Genealogical Register* 17 (1863): 346–54.
Charbonneau, André. *The Fortifications of Île Aux Noix*. Ottawa: Department of Canadian Heritage, 1994.
Charlevoix, Pierre-François-Xavier de. *History and General Description of New France*. 6 vols. Paris, 1744. Trans. and ed. by John Gilmary Shea. Repr., New York: Francis P. Harper, 1900.
Collections of the New-York Historical Society for the Year 1871: Charles Lee Papers. New York: New-York Historical Society, 1872.
Collections of the New-York Historical Society for the Year 1881: Montresor Journals. New York: New York Historical Society, 1882.
De Lery, Chaussegros. "Journal at Carillon, 8 May–2 July 1756." *Fort Ticonderoga Museum Bulletin* 6, no. 4 (1942): 128–44.
Dwight, Nathaniel. "The Journal of Capt. Nathaniel Dwight of Belchertown, Mass., during the Crown Point Expedition, 1755." *New York Genealogical and Biographical Record* 33 (1902): 3–10, 64–70, 164–66.
Fernow, Berthold, ed. *The Records of New Amsterdam from 1653 to 1674*. 7 vols. New York: Knickerbocker Press, 1897.
Fisher, Samuel. Diary, 1758. Manuscript, Library of Congress.
Forbush, Eli. "A Letter from Carillon, (to Rev. Stephen Williams), 4 August, 1759." *Fort Ticonderoga Museum Bulletin* 1, no. 6 (1929): 19–23.
Franquet, Louis. "Voyages et mémoires sur le Canada par Franquet." Institut Canadien de Québec, *Annuaire* 13 (1889).
Furcron, Thomas B. "The Building of Fort Ticonderoga, 1755–1758." *Fort Ticonderoga Museum Bulletin* (1955): 13–67.
Gabriel, Abbé Charles-Nicolas. *Le Maréchal de camp Desandrouins, 1729–1972*. 2 vols. Verdun, France: Renvé-Lallement, 1887.
Gélinas, Cyrille. *The Role of Fort Chambly in the Development of New France, 1665–1760*. Ottawa: Canadian Parks Service, 1983.
Gentleman's Magazine, London.
Gipson, Lawrence Henry. *The British Empire before the American Revolution*. Vol. 5: *Zones of International Friction*. New York: Alfred A. Knopf, 1942.
———. *The British Empire before the American Revolution*. Vol. 6: *The Great War for the Empire, the Years of Defeat, 1754–1757*. New York: Alfred A. Knopf, 1949.

BIBLIOGRAPHY

———. *The British Empire before the American Revolution.* Vol. 7: *The Great War for the Empire, the Victorious Years, 1758–1760.* New York: Alfred A. Knopf, 1946.

Glasier, Benjamin. "French and Indian War Diary of Benjamin Glasier of Ipswich, 1758–1760." *Essex Institute Historical Collections* 86 (1950): 65–92.

Godfrey, William G. *Pursuit of Profit and Preferment in Colonial North America: John Bradstreet's Quest.* Waterloo, Ont.: Wilfrid Laurier University Press, 1982.

Guide to Fort Chambly. Ottawa: Canadian Parks Service, 1928.

Hadden, James. *A Journal Kept in Canada and Upon Burgoyne's Campaign in 1776 and 1777.* Albany, NY: Munsell's Sons, 1884.

Hervey, William. *Journals of the Hon. William Hervey.* Bury St. Edwards, UK: Paul & Mathew, 1906.

Hill, James. "The Dairy of a Private on the First Expedition to Crown Point." *New England Quarterly* 3 (1932): 602–18.

Hunt, George T. *The Wars of the Iroquois: A Study in Intertribal Relations.* Madison: University of Wisconsin Press, 1967.

Jacobs, Wilbur R. "A Message to Fort William Henry: An Incident in the French and Indian War." *Huntington Library Quarterly* 16 (1953): 371–80.

Jameson, Franklin J., ed. *Original Narratives of Early American History: Narratives of the New Netherlands, 1609–1664.* New York: Charles Scribner's Sons, 1909.

Jenks, Samuel. "Journal of Capt. Samuel Jenks." *Proceedings of the Massachusetts Historical Society* 5 (1889): 353–90.

Johnstone, James Chevalier de. *Memoirs of the Chevalier de Johnstone.* 3 vols. Trans. by Charles Winchester. Aberdeen, Scotland: D. Wyllie & Sons, 1870–71.

"A Journal Kept During the Siege of Fort William Henry, August 1757." *Proceedings of the American Philosophical Society* 37 (1898): 143–50.

"Journal of Sergeant David Holden." Proceedings of the Massachusetts Historical Society IV (2nd series). Boston: The Society, 1889, 384–408.

Journal of the Legislative Council of the Colony of New York, Begun the 3rd Day of April 1691 and Ended the 27th of September, 1743. Albany, NY: Weed, Parsons & Co., 1861.

Journal of the Legislative Council of the Colony of New York, Begun the 8th Day of December 1743 and Ended the 3rd of April 1775. Albany, NY: Weed, Parsons & Co., 1861.

Kalm, Peter. *Travels into North America.* Trans. by John Reinold Forster. 3 vols. London: T. Lowndes, 1771.

Kimball, Gertrude Selwyn, ed. *The Correspondence of the Colonial Governors of Rhode Island, 1723–1775.* 2 vols. New York: Houghton, Mifflin & Co., 1902–3.

———. *The Correspondence of William Pitt.* 2 vols. New York: Macmillan Co., 1906.

Knight, Sarah Kemble. *The Journals of Mdm. Knight and Rev. Mr. Buckingham… written in 1704 & 1711.* New York: Wilder & Campbell, 1825.

Knox, John. *An Historical Journal of the Campaigns in North America for the Years 1757, 1758, 1759, and 1760.* 2 vols. London, 1764. Edited by Arthur G. Doughty and reprinted in 3 vols., Freeport, NY: Libraries Press, 1970.

Laramie, Michael. "The French Lake Champlain Fleet and the Contest for the Lake, 1742–1760." *Vermont History* (Winter/Spring 2012): 1–32.

"Les Forts du Régime Français." *Le Bulletin des Recherches Historiques* 24, no. 1 (January 1948): 5–14.

McDonald, de Lery A. "Michel Chartier de Lotbiniere, the Engineer of Carillon." *New York History* 15 (1934): 31–38.

Meech, Susan Spicer, and Susan Billings Meech. *History of the Descendants of Peter Spicer, a Landholder in New London, Connecticut, as Early as 1666.* Boston: Gilson, 1911.

Miller, Reverend John. *A Description of the City and Province of New York, 1695.* London: Thomas Rodd, 1843.

Morgan, William, and Thomas Morgan. *Queen Anne's Canadian Expedition of 1711.* Kingston, Ontario: Jackson Press, 1928.

New American Magazine, Philadelphia.

O'Callaghan, E.B. *History of the New Netherlands; or New York under the Dutch.* 2 vols. New York: D. Appleton & Co., 1855.

———, ed. *Documentary History of New York.* 4 vols. Albany, NY: Weed, Parsons & Co., 1849–51.

———, ed. *Documents Relative to the Colonial History of the State of New York.* 15 vols. Albany, NY: Weed, Parsons & Co., 1856–77.

Paltsits, Victor Hugo. *Minutes of the Executive Council of the Province of New York: Administration of Francis Lovelace, 1668–1673.* 2 vols. Albany, NY: J.B. Lyon Co., 1910.

Pargellis, Stanley M. *Lord Loudoun in North America.* New Haven, CT: Yale Historical Publications, 1933.

———, ed. *Military Affairs in North America, 1748–1765.* New York: D. Appleton–Century Co., Inc., 1936.

Pearson, Jonathan. *A History of the Schenectady Patent.* Albany, NY: Munsell, 1883.

Perley, Sidney, ed. "Dairies Kept by Lemuel Wood, of Boxford..." *Essex Institute Historical Collections* 19 (1882): 61–74, 143–52, 183–92; 20 (1883): 156–60, 198–208, 289–96; 21 (1884): 63–68.

Perry, David. *Recollections of an Old Soldier.* Windsor, VT, 1822. Repr., Cottonport, LA: Polyanthos Press, 1971.

Pomeroy, Seth. *The Journals and Papers of Seth Pomeroy.* Ed. b y Louis Effingham de Forest. New Haven. CT: Tuttle, Morehouse & Taylor Co., 1926.

Pond, Peter. "Experiences in Early Wars in America: Journal of Peter Pond, Born in 1740." *The Journal of American History* 1, no. 1 (1907): 89–93.

Pouchot, Pierre. *Memoirs on the Late War in North America between France and England.* 3 vols. Yverdon, France, 1781. Trans. by Michael Cardy and ed. by Brian Leigh Dunnigan. Youngstown, NY: Old Fort Niagara Association, 1994.

Rocque, Mary Ann. *A Set of Plans and Forts in North America, Reduced from Actual Surveys.* London: M.A. Rocque, 1763.

Rogers, Robert. *The Journals of Major Robert Rogers.* London, 1765. Repr., Ann Arbor, MI: University Microfilms, Inc., 1966.

Roy, Pierre-George. *Hommes et choses du Fort Saint-Fréderic.* Montreal: Les Editions Dix, 1946.

Sainsbury, W. Noel, John W. Fortescue, et al., eds. *Calendar of State Papers, Colonial Series, American and West Indies, Preserved in Her Majesty's Public Records Office.* 45 vols. London: His Majesty's Stationery Office, 1860–1964.

Samuel, Sigmund. *The Seven Years War in Canada, 1756–1763.* Toronto: Ryerson Press, 1934.

A Second Letter to a Friend; Giving a More Particular Narrative of the Defeat of the French Army at Lake George. Boston: Edes & Gill, 1755.

Steele, Ian K. *Warpaths.* New York: Oxford University Press, 1994.

Sullivan, James, and Milton W. Hamilton, eds. *The Papers of Sir William Johnson.* 14 vols. Albany: University of the State of New York, 1921–65.

Thwaites, R.G., ed. *The Jesuit Relations and Allied Documents.* 73 vols. Cleveland, OH: Burrow Bros. Co., 1896–1901.

Tomlinson, Abraham, ed. *The Military Journals of Two Private Soldiers, 1758–1775.* Poughkeepsie, NY: A. Tomlinson, 1855.

Vauban, Sébastian Le Prestre de. *A Manual of Siegecraft and Fortification.* Ann Arbor: University of Michigan Press, 1968. Trans. and ed. from the 1740 edition by George A. Rothrock.

Vaudreuil, Marquis de. "Marquis Vaudreuil to Sieur de Lotbiniere, 20 September 1755." *Fort Ticonderoga Museum Bulletin* 1, no. 3 (1928): 2–3.

Verney, Jack. *The Good Regiment*. Montreal: McGill-Queen's University Press, 1991.
"Wadsworth's Journal." *Massachusetts Historical Society Collections*, ser. 4, no. 1 (1852): 105–6.
Webster, Clarence J., ed. *The Journals of Jeffery Amherst*. Toronto: Ryerson Press, 1931.
Westbrook, Nicholas, ed. "Like Roaring Lions Breaking from Their Chains." *The Bulletin of the Fort Ticonderoga Museum* 16, no. 1 (1998): 16–91.
Wright, John W. "Sieges and Customs of War at the Opening of the Eighteenth Century." *American Historical Review* 39, no. 4 (1934): 629–41.

Index

A

Abercromby, James 13, 15, 60, 61, 62, 63
Artillery Cove 93

B

Beaucours, Josue Boisberthelot de 29
Beauharnois, Charles 31, 42, 43, 45, 47
Bigot, François 33
Binckes, Jacob 102
Bougainville, Louis-Antoine 38, 39, 41
Bourdon, Jean 24
Bourlamaque, Charles 37, 38, 39, 65, 68

C

Carignan-Salières Regiment 24
Chambly Basin 29
Chambly, Jacques de 24, 25, 26
Chimney Point 42, 45
Connecticut River 99
Corne, Jean-Louis de la 42, 43
Crown Point 42, 45

D

Denonville, Marquis de (Jacques-Rene de Brisay) 42
Desandrouins, Jean-Nicolas 68, 93
Diable 39
d'Olabaratz, Jean 39

E

Evertsen, Cornelis 102

INDEX

Eyre, William 69, 85, 87, 88, 89, 90, 93

F

Fitch, Thomas 89
Fort Albany 80, 103
Fort Amsterdam 99, 100, 101
Fort Anne 82
Fort Carillon 57, 59
Fort Chambly 9, 28, 31, 41, 42, 47, 80
Fort Clinton 83
Fort Crosby 83
Fort Edward 89, 90, 92
Fort Frederick 83
Fort George 134
Fort Ingoldsby 80
Fort James 102
Fort Nassau 73
Fort Orange 74, 100
Fort Richelieu 21, 22, 24, 25, 28
Fort Schuyler 80
Fort St. Frederic 31, 33, 34, 50, 53, 56, 68
Fort St. Jean 33, 35, 36, 57
Fort St. Louis 24, 25, 26, 28
Fort St. Therese 28, 33
Fort Vaudreuil 53, 54, 57

G

Guinea 100

H

Half Moon 79, 80
Hardy, Charles 89
Hebecourt, Louis-Philippe 67
Hocquart, Gilles 31, 45, 47

I

Île aux Noix 16, 17, 20, 35, 37, 41, 65, 72

K

Kalm, Peter 33, 49

L

La Chute River 54, 57, 63
La Prairie 33
Lery, Chaussegros de 31, 33, 45, 47, 48
Leverett, John 99
Levis, Gaston 38, 59
Lotbiniere, Michel 38, 39, 51, 52, 53, 54, 56, 57, 58, 59, 68
Louis 26
Louisbourg 15
Lovelace, Francis 102, 103
Lusignan, Paul-Louis 17, 41, 57

M

Manning, John 103, 105
Missisquoi Bay 37

INDEX

Montcalm, Marquis de 16, 20, 38, 59, 65, 93, 95
Montmagny, Charles Hualt de 22
Morandière, Rocbert de la 43

N

Nicolls, Richard 100, 101, 102
Nustigione 79

O

Oswego 43

P

Point a Margot 35
Pointe-à-la-Chevelure 42, 43, 45, 47

R

Rabelin, Jacques 23
Richelieu River 21, 24, 31, 33, 35
Riviere du Sud 37
Rogers Island 95

S

Salières, Marquis de (Henri de Chastelard) 25, 26
Saratoga 83
Saurel, Pierre de 24, 26
Schenectady 83

Sedgwick, Robert 99
St. Jean 39
St. Lawrence River 22, 43
Stuyvesant, Peter 99, 100, 101, 103

T

Ticonderoga 51, 54, 56, 133, 134
Treaty of Utrecht 45

V

Vaudreuil, Phillip 29, 31, 37, 53
Vigilante 39

W

Waldegrave, Earl of 45
Whiting, Nathan 92
Williamson, Adam 94
Wood Creek 42, 80
Wooster, David 63

About the Author

Michael G. Laramie is the author of *The European Invasion of North America: Colonial Conflict Along the Hudson-Champlain Corridor, 1609–1760*; *By Wind and Iron: Naval Campaigns in the Champlain Valley, 1665–1815* (Westholme, 2014); and *King William's War: The First Contest for North America, 1689–1697* (Westholme, 2017). He lives with his family in Arizona.

Visit us at
www.historypress.com

www.ingramcontent.com/pod-product-compliance
Lightning Source LLC
Chambersburg PA
CBHW042142160426
43201CB00022B/2382